D0845635

physical sciences data 2

statistical treatment of experimental data

physical sciences data 2

statistical treatment of experimental data

j.r. green, b.sc., ph.d.

Lecturer in Computational and Statistical Science, University of Liverpool

d. margerison, b.sc., ph.d.

Senior Lecturer in Inorganic, Physical and Industrial Chemistry, University of Liverpool

ELSEVIER SCIENTIFIC PUBLISHING COMPANY
Amsterdam – Oxford – New York 1978

ELSEVIER SCIENTIFIC PUBLISHING COMPANY
335 Jan van Galenstraat
P.O. Box 211, 1000 AE Amsterdam, The Netherlands

Distributors for the United States and Canada:

ELSEVIER NORTH-HOLLAND INC.
52, Vanderbilt Avenue
New York, N.Y. 10017

First published 1977
First revised reprint 1978

Library of Congress Cataloging in Publication Data

Green, John Robert.
Statistical treatment of experimental data.

(Physical sciences data ; 2)
Bibliography: p.
Includes index.
1. Mathematical statistics. 2. Science--Methodology.
I. Margerison, D., joint author. II. Title.
III. Series.
QA276.G715 1978 519.5 78-18263
ISBN 0-444-41725-7

ISBN 0-444-41725-7 (Vol.2)
ISBN 0-444-41689-7 (Series)

Printed in The Netherlands

PREFACE

The authors have both found the task of writing this book very instructive and rewarding in itself. Since basically one of us is a statistician and the other an experimentalist, this has proved to be a useful blend for writing such a book and each of us has learned from the other. We hope the readers will find the resulting book as instructive and profitable as we did in writing it.

Our grateful thanks are due to the typists of our two departments who typed various drafts and especially to Mrs. Jean Powell who carefully typed the final version for photographic reproduction. We also wish to express our appreciation to Mr. A.J. Nicholson and Mr. M.H. Carruthers for help in various ways and especially to Dr. D.L. Dare and Mr. J.M. Gorman for providing some of the experimental results. We are grateful also to Professors M.R. Sampford and C.H. Bamford and other colleagues for their interest and comments, and not least to the publishers for inviting us to write this book. Finally, we must express our indebtedness to our wives for their help and forbearance.

J.R. Green

D. Margerison

PREFACE TO THE FIRST REVISED REPRINT

The early necessity for a second printing has been gratifying as it supports the authors' original belief that the book meets a real need. The opportunity has been taken to correct a few minor errors and we are grateful to those readers who have pointed some of these out.

J.R. Green

D. Margerison

CONTENTS

CHAPTER 1

INTRODUCTION

This book is primarily intended for experimentalists, particularly those who work in the physical sciences. It deals with some of the common statistical methods which may be employed to treat experimental data. As we shall see in subsequent chapters, any statistical analysis of data has to be based upon some assumed probability model for those data. This is an aspect of statistical inference which is frequently glossed over in elementary texts which often deal almost exclusively with computational procedures. In contrast, we have devoted a good deal of space to the ideas and reasoning behind statistical methodology. The gain in understanding and confidence to be obtained from a formal structuring of the problems posed in data treatment far outweighs the initial difficulty of grasping new concepts and notation. Throughout the book, we have illustrated most of our formal results with numerical examples usually taken from the laboratory. We expect that the examples we have chosen will serve as models for the treatment of the problems of our readers.

We commence in Chapter 2 by dealing briefly with the basic notions of probability since these lie at the centre of statistical reasoning. In Chapter 3 we further develop probability theory, defining important concepts such as random variable, density function, distribution function, expectation, mean and variance. Chapter 4 deals with some important probability distributions. Much of it is concerned with consolidating and exemplifying the material of Chapter 3, while at the same time preparing the ground for later chapters.

In Chapter 5, we consider the problem of estimation of the unknown parameters of a probability distribution from observations we make in a limited number of experiments. Chapter 6 then explains how we may compute a confidence interval for such a parameter. The ideas and concepts of hypothesis testing are considered in a general way in Chapter 7 which will be frequently called upon in the remainder of the book. Chapters 8, 9, and 10, for example, describe tests on means, variances, and goodness of fit.

Chapter 11, dealing with correlation, paves the way for our discussion of the straight line and the polynomial. We have given an extensive discussion of these matters in Chapters 12 to 15. Chapters 12 and 13 deal with the situation where the straight line or polynomial is constrained to pass through the origin or some other fixed point while the remaining two chapters discuss the more general cases. Finally in Chapter 16, we draw together the main results of Chapters 12 to 15 in a more general treatment. This last chapter is likely to prove difficult to many of our readers and can be omitted without serious loss if they are content to deal only with the restricted data models of Chapters 12 to 15.

We believe that it will be useful to draw attention to those topics which we have dealt with which are not usually found together in a statistical book at this level. These include a discussion of rounding-off errors (Chapters 3 and 5) and the choice of the number of significant figures (Chapter 5). We have also dealt with the estimation of the mean and variance of a function, including the case where the estimates of variance of the various arguments are associated with different numbers of degrees of freedom (Chapter 5). In the chapters on testing, we have considered testing for homogeneity of variance (Chapter 9), stabilization of variance (Chapter 9), and testing for outliers (Chapter 10). In Chapter 8 we give tests for deciding which of a set of significantly different means are significantly different from which. Also included in Chapters 8 and 9 is a brief discussion of what estimates to use for the various parameters when the test is completed. Chapters 12 to 15 contain several unusual features. For a start, the straight line or polynomial constrained to pass through the origin or some other fixed point is seldom treated at length, despite the frequency of occurrence of this type of relationship. In Chapter 13, we extend the idea of data transformation, already mentioned in the earlier chapters as a means of simplifying arithmetic, to include orthogonal polynomials in x. These functions of x greatly simplify the algebra of polynomial fitting and give quick and neat solutions to the problem of the general straight line. For the most part, we have given a general treatment involving the use of weights throughout these four chapters; weights need to be used when the random variable, Y, has a variance dependent on the value of x. Other unusual features of these chapters are the methods for inverse interpolation when using straight lines, and testing for homogeneity of slope and/or intercept, in Chapters 12 and 14; we also give the corresponding tests for coincidence of a set of polynomial regressions in Chapters 13 and 15.

Having said what we have tried to do, we have to emphasize that the method of acquisition of the data is just as important as the analysis of the data. This is because any data analysis has to be based on certain assumptions, in fact on a model of the data. Obviously we can do little more here than indicate some general considerations which should be borne in mind in planning a series of experiments.

Firstly, there must be a clearly defined objective. This is not crucial for a preliminary, exploratory, experiment, but, for experiments generating data to test hypotheses, these hypotheses must be formulated in advance, not on the basis of the data to be used for tests. It is usually useful to write down an assumed model for the data, as will be amply illustrated in this book. Secondly, we must avoid systematic bias, so that each of our observations will be actually equal to a supposed quantity, μ, plus random error (which is usually equally likely to be positive or negative). Bias is easily unwittingly introduced. For example, we may need to randomize the order of our observations to avoid a bias due to a time-trend in observed values.

Thirdly, we should design our experiment so that we will be able to obtain a measure of the variability of the random error of our observations to enable us to distinguish between real systematic differences and those due simply to random fluctuation. Fourthly, our use of equipment and design of experiment (including numbers of observations) should be arranged efficiently, that is, so as to minimise random error and to accentuate the sizes of systematic differences relative to that random error.

Finally, we should ensure that our observations are really representative of the population, or source of such data, to which we would like to apply our conclusions. It is all too easy to actually draw our observations from a narrower, less representative, source.

The general pattern of the book will now be becoming clear. We assume that the experimentalist has some clear model of his data, in part perhaps from theory, in part from his knowledge of the way his apparatus functions. We shall assume that he has made sufficient observations for his purpose, a number which will become clearer as the book proceeds. Given this background we shall describe what he can do with his data, what assumptions he will need to be able to make, and the extent to which his data enable him to draw useful conclusions from his experiments.

We have not provided statistical tables in this book since the main ones are widely available, and paperbacks of statistical tables can be purchased quite cheaply. Also we have given references to well-known collections for the more specialized tables if these are required.

CHAPTER 2

PROBABILITY

In a deterministic experiment, the known conditions of the experiment fully determine the outcome. We shall not be concerned with such a rare kind of experiment. Usually some factors are not fully controlled and certainly not fully known to the experimenter. We describe such experiments as random experiments. To deal with them properly, we need to know some of the basic ideas of probability theory. This chapter deals with these ideas.

1. Some basic definitions

The set of all possible outcomes from a random experiment is called the sample space. We may think of the outcomes as points in the sample space. Two examples are given below.

(i) In a certain experiment, the number of radioactive particles emitted from a source during a certain time interval is counted. The sample space consists of the numbers 0,1,2,...

(ii) In another experiment, the volume of a solution that must be added to a substance in a flask to produce a certain colour change is measured. The sample space consists of the positive half of the real line, $x > 0$.

These two examples also illustrate the terms discrete in (i), that is where there are a countable number of possible outcomes (which may be infinite as in this particular example), and continuous in (ii), that is where all possible real values in a certain interval (finite or infinite), or series of intervals, may occur.

In each of the two examples, (i) and (ii), only one number is recorded, that is both these experiments are univariate. If more than one value is obtained from a single performance of an experiment, it is multivariate; if two values are obtained, for example, it is bivariate.

We shall call a set of outcomes in the sample space an event and say that the event occurs if any one of the outcomes which comprise it occurs when the experiment is performed.

We may think of the probability of an event as a degree of belief, or expectancy, that the event will occur, but this is not essential to its definition. The probability of an event A, say, is a number P(A) satisfying certain conditions. Let B be any other event which has no outcome in common with A, that is A and B are disjoint sets of points. The conditions which must be satisfied by the numbers P(A) and P(B) are:

$$\text{(i)} \quad 0 \leq P(A) \ ,$$

$$\text{(ii)} \quad P(A \cup B) = P(A) + P(B) \ ,$$

$$\text{(iii)} \quad P(S) = 1 \ ,$$

where $A \cup B$ represents the union of the two sets, the set of all points belonging to A or B, and S stands for the whole sample space. Axiom (ii) is sometimes called the addition law of probability. Basically these axioms are all we need, but it is convenient to define and use other concepts.

We may easily deduce from these axioms that $P(A) \leq 1$, and also that

$$P(\bar{A}) = 1 - P(A) \ , \qquad (1)$$

where $\bar{A}$, read as 'A bar' or 'not A', is the set of all points of S not in A.

Clearly axiom (ii) extends to any number of mutually exclusive events, thus

$$P(A \cup B \cup C \ldots) = P(A) + P(B) + P(C) + \ldots \qquad (2)$$

If two events, say D and E, are not mutually exclusive, they may nevertheless be split up into three mutually exclusive events: $D - D \cap E$, $E - D \cap E$, and $D \cap E$, as illustrated in the "Venn diagram", Figure 2.1. Here $D \cap E$ represents the set of points common to both D and E.

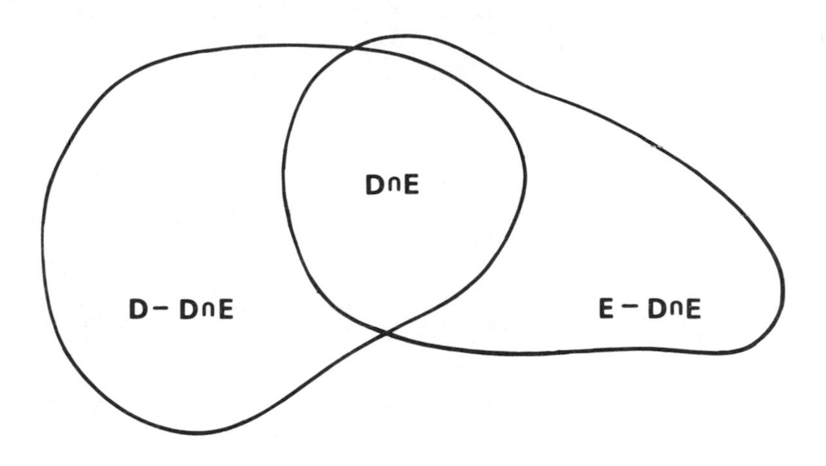

Figure 2.1. A Venn diagram illustrating equation (3)

Now the union of these three sets is clearly D∪E and hence from equation (2)

$$P(D \cup E) = P(D-D \cap E) + P(E-D \cap E) + P(D \cap E)$$

$$= P(D) + P(E) - P(D \cap E) . \quad (3)$$

Equation (3) clearly degenerates to axiom (ii) if D and E are mutually exclusive. It can be similarly shown that quite generally for events F, G, H

$$P(F \cup G \cup H) = P(F) + P(G) + P(H) - P(F \cap G) - P(G \cap H) - P(F \cap H) + P(F \cap G \cap H) , \quad (4)$$

etc.

2. Independence and conditional probability

Any two events, A and B, are called independent if the probability that both occur, $P(A \cap B)$, equals the product of the probability of A and that of B, viz.

$$P(A \cap B) = P(A).P(B) \; . \qquad (1)$$

This is often called the product law of probability.

More generally, the conditional probability of any event C, given that an event D has occurred, written as $P(C|D)$, where the vertical stroke is read as 'given', is defined by the equation

$$P(C|D) = P(C \cap D)/P(D) \; . \qquad (2)$$

Hence

$$P(C \cap D) = P(D).P(C|D) \qquad (3)$$

and

$$= P(C).P(D|C) \text{ by symmetry } . \qquad (4)$$

For example, suppose we have a bag containing three red and four blue discs. Then, if we draw discs randomly (that is, each disc contained in the bag has an equal probability of selection) without replacement, the probability of drawing blue the second time, given red was drawn the first time, is $4/6 = 2/3$. The probability of drawing one red and then one blue disc is given as

$$P(R \text{ then } B) = P(R \text{ first}).P(B|R)$$

$$= \frac{3}{7} \times \frac{2}{3} = \frac{2}{7} \; .$$

This is also equal to $P(B \text{ first}).P(R|B)$ as shown below

$$P(B \text{ first}).P(R|B) = \frac{4}{7} \times \frac{3}{6} = \frac{2}{7} \; .$$

$P(R \cap B)$ is the sum of these two probabilities, namely $4/7$.

By a simple extension of equation (3), we obtain the general result

$$P(C \cap D \cap E \cap F \ldots) = P(C).P(D|C).P(E|C \cap D).P(F|C \cap D \cap E) \ldots \quad (5)$$

In the same way as for events, two or more performances of experiments are called _independent_ if the probabilities of the different outcomes in any one performance are unaffected by the other outcomes. Independent repeat performances of an experiment are usually called _replicates_.

It can be proved that the proportion of times any event A occurs in n replicates of an experiment tends (in a probabalistic sense) to P(A) as n tends to infinity, so that for large n this proportion is a good approximation to P(A). This gives us another understanding of the meaning of probability as a long-run proportion and reveals a stabilizing effect in the midst of randomness.

3. The discrete uniform probability model

An example of a simple probability model is the discrete uniform model, where each outcome is equally likely. If there are m different outcomes, then the probability of each is 1/m, and, for any event A

$$P(A) = p/m , \quad (1)$$

where p is the number of outcomes in A. For example, tossing an unbiased (fair) die, P(i) = 1/6, i = 1,2,...6, and P(1 or 2) = 1/3.

A more complicated example relates to drawing a random sample of size s from a batch of size N. By the term random sample, we mean that all samples of size s are equally likely to be chosen. The number of possible samples is $^{N}C_{s}$ (the number of combinations of s out of N things which is $N!/\{s!(N-s)!\}$ and so the probability of choosing a particular one is $1/^{N}C_{s}$. If, now, r of the N items are 'specials', different in some way from the rest, the number of ways of drawing a sample containing d specials is the number of ways of choosing s items such that d are specials and s-d are non-specials, that is ${}^{r}C_{d} \times {}^{N-r}C_{s-d}$. Hence the probability of drawing a sample containing d specials is given by equation (1) as

$$P(d \text{ specials}) = {}^{r}C_{d}.{}^{N-r}C_{s-d}/{}^{N}C_{s} \ , \ d = 1,2,\ldots \min(r,s) \ ,$$

where min(r,s) is the smaller of r and s. This actually defines the hypergeometric distribution, but it is here thought of simply as a special case of the uniform model.

We may use the above result to obtain the probability of drawing a red and blue disc from a bag containing three red and four blue discs without replacement - the same problem that we discussed earlier. Here we draw a random sample of size 2 (= s) from a batch of size 7 (= N), with 3 (= r) of the batch as 'specials' (red, in this case), and we wish to calculate the probability that 1 (= d) of our sample is a special (a red). From the above expression,

$$P(R \cap B) \quad = \quad {}^{3}C_{1}\,{}^{4}C_{1}/{}^{7}C_{2} \ .$$

Thus

$$P(R \cap B) \quad = \quad 3 \times 4/21 = 4/7 \ ,$$

the same result as we obtained earlier.

This example illustrates the fact that the hypergeometric distribution relates to sampling without replacement.

BUTIONS

s performed,
or to a con-
ete, we may
possible out-
le event. How-
y think of our
and y + dy.
interval as
tually in
ns, because we
decimal places,
acent possible
oximation to
tinuous distri-
an infinite
ituation, we
3.5, so that
e probability
h an event is
an almost
alled an almost

ple space, that
variable takes a
of it. The set
prises its
e when its
s sample space
f using capital
for realiza-
e, P(X = x),
ability of the
event comprising all outcomes for which X takes the value x.
The corresponding quantity in the case of a continuous random

variable is $P(x < X \leq x+dx)$, the probability of the event comprising all outcomes for which X falls in the interval $(x, x+dx)$.

A function of a random variable is itself a random variable. Any functional relationship between random variables applies equally to their realizations.

The simplest case, if the outcomes are numerical values, is where X takes the value of the outcome itself. For example, the sample space for the random experiment of tossing a die is

$$S = \{1, 2, 3, 4, 5, 6\} .$$

One random variable, X, which we might define on this sample space is one whose realizations are the elements of S. Thus, if the die is fair,

$$P(X = 1) = P(1) = 1/6 .$$

It is important to appreciate that the realizations of a random variable X need not be the outcomes in the sample space S for the experiment. For example, we might define a random variable X for the random experiment of tossing a die which takes the value 0 when the outcome is even and the value 1 when the outcome is odd. Thus if the die is fair,

$$P(X = 1) = P(1) = \tfrac{1}{2} .$$

These two examples illustrate that it is necessary to define carefully exactly how the realizations of a random variable are related to the sample space for the experiment.

Alternative names for a random variable are <u>statistic</u> or <u>variate</u>.

1. Density function

It will be clear that the possible realizations of X form a set which we shall now represent by S.

If a random variable X is continuous, we may specify a <u>probability density</u> $f(x)$, which is such that the integral of

$f(x)$ over any interval A gives the probability that X belongs to A, $P(X \varepsilon A)$, or, more simply, the probability of A, $P(A)$. That is,

$$P(X \varepsilon A) = P(A) = \int_A f(x)\,dx \,.$$

The probability of the simple event that X belongs to the interval $(x, x+dx)$ is $f(x)dx$. Here, $f(x) \geq 0$, all x, and

$$\int_{-\infty}^{+\infty} f(x)\,dx = 1 \,.$$

Often $f(x)$ is defined over the whole real line, taking the value 0 where X cannot occur.

The function f is usually called the (probability) density function.

In the discrete case, a single point (or single possible outcome), x, is a simple event and we shall express its probability as $p(x)$. Then, as before,

$$P(X \varepsilon A) = P(A) = \sum_{x \varepsilon A} p(x) \,,$$

the summation, as indicated by the subscript, being taken over the elements of A. Here $1 \geq p(x) \geq 0$, all x, and

$$\sum_{x \varepsilon S} p(x) = 1.$$

In a bivariate situation where two different outcomes are recorded for each performance of an experiment, the two corresponding random variables, say X and Y, have, if they are continuous, a joint density $f(x,y)$, which is such that

$$f(x,y) \geq 0 \text{ for all } x \text{ and } y \,,$$

$$P(x < X \leq x+dx \text{ and } y < Y \leq y+dy) = f(x,y)\,dxdy \,,$$

$$P(X \varepsilon A \text{ and } Y \varepsilon B) = \int_A \int_B f(x,y)\,dxdy \text{ for all sets A}$$

and B, and $\int_{-\infty}^{+\infty} \int_{-\infty}^{+\infty} f(x,y)\,dxdy = 1 \,.$

This exemplifies a multivariate joint density function, $f(x,y,z,\ldots)$, which has corresponding properties.

The two random variables, X and Y, are called <u>independent</u> if

$$P(X\varepsilon A \text{ and } Y\varepsilon B) = P(X\varepsilon A).P(Y\varepsilon B) \text{ for all sets A and B .}$$

If and only if X and Y are independent, their joint density factorizes into a function of x and a function of y, thus

$$f(x,y) = f_1(x).f_2(y) ,$$

where f_1 and f_2 are the density functions of X and Y, called <u>marginal</u> density functions. Marginal density functions also exist for non-independent random variables but we shall not consider them here.

Similar definitions may be applied to the case where we have two or more discrete random variables.

<u>Example</u>. We may illustrate the use of a continuous density function by considering errors introduced in rounding-off. Figure 3.1 illustrates a typical situation in which this type of error is of importance. The figure shows two marks inscribed on a steel rod and a steel rule calibrated in cm and mm placed alongside the rod for the purpose of measuring the distance between the two marks.

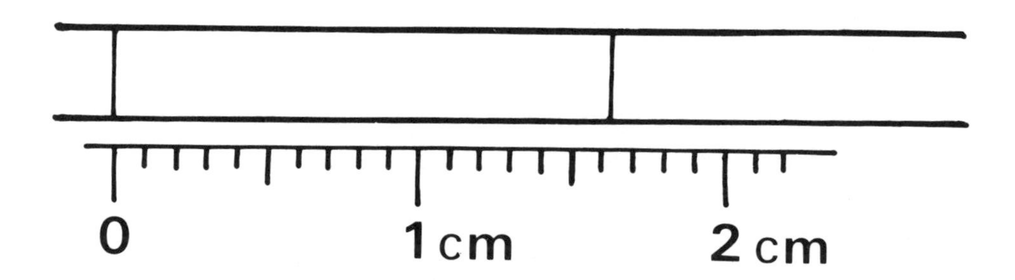

Figure 3.1. Measurement of a distance

We decide to read the rule to the nearest mm thereby introducing a rounding-off error somewhere in the range ±0.5 mm. Thus, in the example in the figure, we read the distance between the two marks as 16 mm though the true distance is clearly something greater than this. Now another person seeing the recorded value of 16 mm only knows that the true value lies in the range (16±0.5) mm. The problem we set ourselves here is to calculate the probability that the magnitude of error is greater than 0.35 mm.

To solve the problem, let us define a random variable X whose realizations are

$$x = (d_{true} - d_{obs})/\text{mm} ,$$

where d_{true} and d_{obs} are the true and observed separations of the two marks. Let us assume that X has a continuous uniform probability distribution over the interval ±0.5. In other words, it is just as likely for the error to fall in an interval dx around 0.1 as in an interval dx around, say, 0.2. Figure 3.2 illustrates the density function for this particular case.

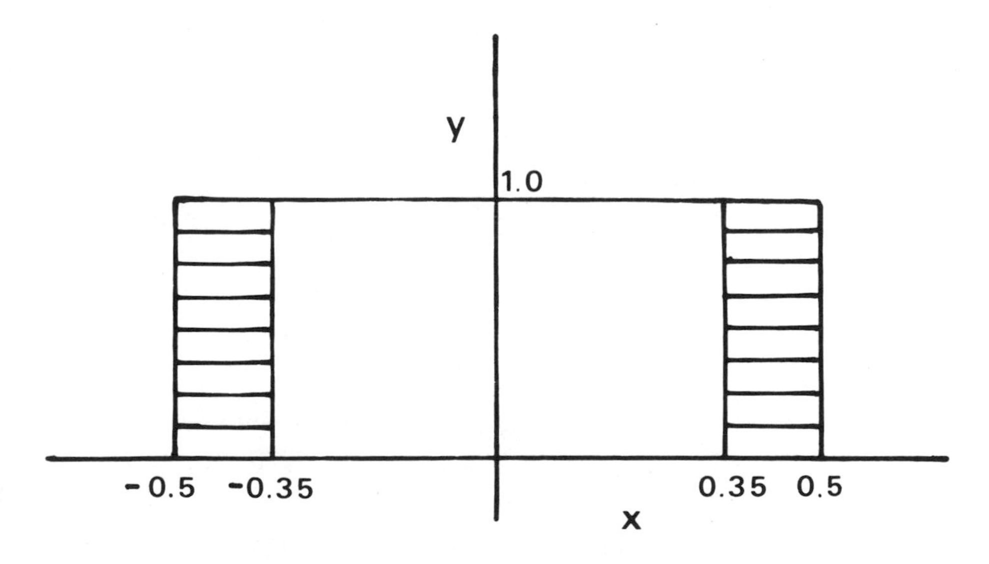

Figure 3.2. y = f(x) , the density function for a rectangular distribution.

Here the density is 1 in the range (-0.5,0.5) since the total area under the figure is 1. To obtain the answer to our question we require to integrate the density over (-0.5,-0.35) and (0.35,0.5) which gives 0.3 as the required probability.

A continuous uniform distribution such as that illustrated in Figure 3.2 is usually called a rectangular distribution since the graph of f(x) against x is rectangular if the vertical walls at the ends are included.

2. Distribution function

The distribution function of a random variable X is defined to be

$$F(x) = P(X \leq x) = \begin{cases} \sum_{y \leq x} p(y) , & \text{if X is discrete} \\ \int_{-\infty}^{x} f(y)dy , & \text{if X is continuous} \end{cases} \tag{1}$$

Here y stands for a realization of X. It follows from equation (1) that

$$P(a < X \leq b) = F(b) - F(a) . \tag{2}$$

We note that as $a \rightarrow b$, $P(a < X \leq b) \rightarrow 0$ for continuous X, so $P(X = b) = 0$, for any b, so that X = b illustrates an 'almost impossible' event as already mentioned. Since the probability of the equality X = b is zero for a continuous distribution, it follows that the inequalities $<$ and $\leq$ (or $>$ and $\geq$) in such expressions as $P(a < X \leq b)$ are equivalent.

Note that statisticians use the term, distribution function, in a different sense to that often used by chemists and physicists, who frequently apply the term to the probability density. The density and distribution functions are clearly distinguished in Figures 3.3 and 3.4 for the case of the normal distribution (to which we shall refer extensively in later chapters). These figures also show the β quantile, ξ_β which is such that $F(\xi_\beta) = \beta$, where $0 \leq \beta \leq 1$. This quantile is also called the 100β percentile. For instance, $\xi_{0.1}$, the 0.1 quantile, is such that the probability of observing a value of X less than or equal to $\xi_{0.1}$ is 0.1.

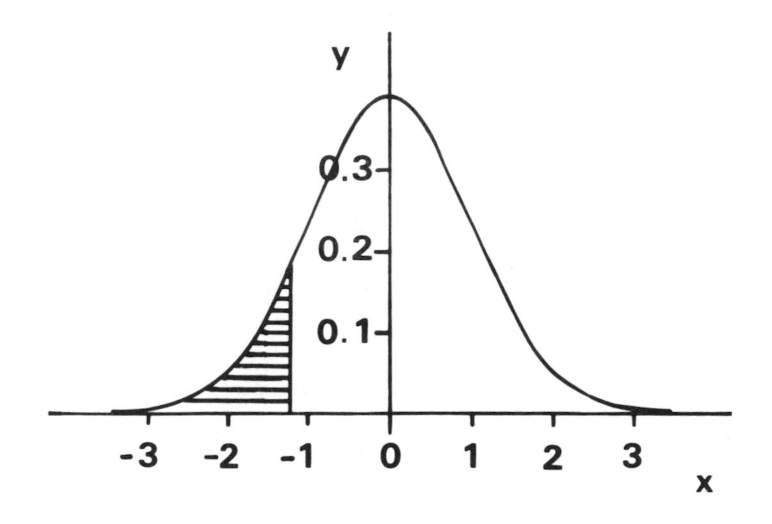

Figure 3.3. $y = f(x)$, the standard normal density function. The shaded area is β up to $x = \xi_\beta$.

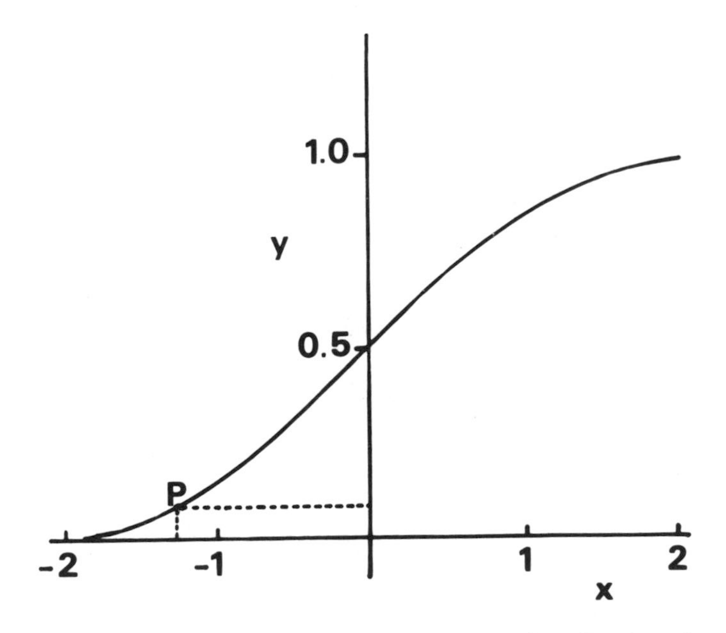

Figure 3.4. $y = F(x)$, the standard normal distribution function. The point P is (ξ_β, β).

3. Expectation

The expectation of any function g of a random variable X is defined to be

$$E\{g(X)\} = \begin{cases} \Sigma g(x)p(x) \quad , \text{ if X is discrete} \\ \int_{-\infty}^{+\infty} g(x)f(x)dx \quad , \text{ if X is continuous .} \end{cases} \tag{1}$$

These two forms are both included in the Steltjes integral form

$$E\{g(x)\} = \int_0^1 g(x)dF(x) \ , \tag{2}$$

which we may conveniently think of as representing whichever form of the right-hand side of equation (1) is appropriate. Some particularly important cases are defined below.

(i) The distribution mean of X (also called the mean of the distribution F(x))

This mean is often represented by the Greek letter μ and is defined as

$$\mu = E(X) = \int_0^1 xdF(x) \ . \tag{3}$$

The mean affords a measure of central location of the distribution.

(ii) The rth (non-central) moment of X (or of the distribution F(x))

This is often designated μ_r' and is defined by

$$\mu_r' = E(X^r) = \int_0^1 x^r dF(x) \ , \ (r = 1,2,...) \tag{4}$$

provided that the integral is finite. Clearly

$$\mu_1' = \mu \ .$$

(iii) The rth central moment of X (or of F(x))

This quantity is often designated by the symbol μ_r and is defined by

$$\mu_r = E\{(X-\mu)^r\} = \int_0^1 (x-\mu)^r dF(x) \quad ,(r = 1,2,\ldots) \tag{5}$$

provided the integral is finite. As before, μ here stands for E(X). When the distribution is symmetrical about μ, all the odd central moments are zero.

(iv) The distribution variance of X (or of F(x))

This is the second central moment, that is μ_2. More commonly, the Greek letter σ^2, or sometimes $\sigma^2(X)$, is used for this quantity, which is defined by

$$\sigma^2 = E\{(X-\mu)^2\} = \int_0^1 (x-\mu)^2 dF(x) \ . \tag{6}$$

We shall also frequently represent the variance of X by the symbol V(X):

$$V(X) = \sigma^2 = \mu_2 \ .$$

The square root of the (distribution) variance is the (distribution) standard deviation, σ. Both σ^2 and σ afford measures of the spread of the distribution about the mean, σ being in the same units as X itself.

(v) The moment generating function or m.g.f.

This is often represented by the symbol $M_X(t)$ where t is a dummy variable. It is defined through the expression

$$M_X(t) = E\left(e^{tX}\right) , \qquad (t \geq 0) \ , \tag{7}$$

when it exists for some interval of t including 0. If moments of X exist up to the rth, the m.g.f. can be expanded for small t , as

$$M_X(t) = 1 + \mu_1't + \frac{\mu_2't^2}{2} + \ldots \frac{\mu_r't^r}{r'} + o(t^r) \;, \tag{8}$$

where o means 'small compared with.'

The moment generating function is <u>unique</u>; that is, if two random variables have the same m.g.f. they have the same probability distribution.

For two random variables X and Y, the m.g.f. of X and Y is

$$M_{X,Y}(t_1,t_2) = E\{\exp(t_1X + t_2Y)\} \;. \tag{9}$$

If and only if X and Y are independent, this factorises into the product of the m.g.f. of X and that of Y — similarly for more than two random variables.

(vi) <u>The characteristic function, c.f., and the probability generating function, p.g.f.</u>

These two functions are closely related to the m.g.f. The characteristic function, which always exists, is defined by

$$\text{c.f.} = E\left(e^{itX}\right) \;, \tag{10}$$

where $i = \sqrt{-1}$.

The probability generating function is defined by

$$\text{p.g.f.} = E\left(t^X\right) \tag{11}$$

$$= \sum_x p(x)t^x \text{ for a discrete random variable.}$$

Writing g(t) for the p.g.f., we obtain

$$g'(t) = E\left\{Xt^{X-1}\right\}$$

and $$g''(t) = E\left\{X(X-1)t^{X-2}\right\} \;.$$

Hence

$$g'(1) = E(X) = \mu \tag{12}$$

and $$g''(1) = E\{X(X-1)\} = E(X^2) - \mu . \tag{13}$$

Equations (12) and (13) can be useful for deriving the mean and variance of a probability distribution (see Chapter 4).

We now designate the m.g.f., the c.f. and the p.g.f. by $g_1(t)$, $g_2(t)$, and $g_3(t)$ respectively. Where it is necessary to specify the random variable concerned, we shall write this as a second argument.

Thus

$$g_1(t) = g_1(t,X) = E\left(e^{tX}\right) ,$$

$$g_2(t) = g_2(t,X) = E\left(e^{itX}\right) ,$$

$$g_3(t) = g_3(t,X) = E\left(t^{X}\right) .$$

From these expressions, it may readily be demonstrated that

$$g_1(it) = g_2(t) \text{ and } g_2(-it) = g_1(t) ,$$

$$g_1(\ln t) = g_3(t) \text{ and } g_3(e^t) = g_1(t) ,$$

$$g_2(-i\ln t) = g_3(t) \text{ and } g_3(e^{it}) = g_2(t) ,$$

where these exist. This shows us how to derive any one of these generating functions from another. Further,

$$g_1(t,aX + b) = e^{bt}g_1(at,X) ,$$

$$g_2(t,aX + b) = e^{ibt}g_2(at,X) ,$$

$$g_3(t,aX + b) = t^{b}g_3(t,aX) ,$$

where a and b are any constants. This shows us the effects of any linear transformation on the generating functions. Moreover if X and Y are independent random variables

$$g_i(t,X+Y) = g_i(t,X).g_i(t,Y), \quad \text{for } i = 1,2,3.$$

(vii) General remarks about the above and other generating functions

There are other generating functions of interest and usefulness, such as the cumulant generating function (c.g.f.) and the factorial moment generating function (f.m.g.f.). However, we shall not discuss these further except to say that for these and each of the other generating functions we have mentioned, there is a uniqueness theorem: if two random variables have the same generating function (m.g.f. for instance), then they have the same probability distribution.

When the generating function is the expectation of a function of the random variable (as is true for the m.g.f., the c.f., and the p.g.f., for instance), the generating function of the sum of two independent random variables is equal to the product of the two generating functions for the separate random variables. Among other things, this property, together with the corresponding uniqueness theorem, is useful for deriving additivity properties for certain distributions, as we shall see exemplified in Chapter 4.

(viii) The covariance of two random variables X and Y

If two random variables X and Y have means μ_X and μ_Y respectively, the covariance of X and Y, represented by C(X,Y), is defined to be

$$C(X,Y) \quad = \quad E\{(X-\mu_X)(Y-\mu_Y)\} \ . \tag{14}$$

We observe that the quantity, variance, is a special case of covariance, viz.

$$C(X,X) \quad = \quad V(X) = E\{(X-\mu_X)^2\} \ .$$

An associated quantity, the distribution correlation coefficient, is defined by the expression

$$\rho(X,Y) \quad = \quad \frac{C(X,Y)}{\{V(X).V(Y)\}^{\frac{1}{2}}} \ . \tag{15}$$

When no ambiguity exists we shall contract $\rho(X,Y)$ to ρ. The value of ρ lies in the range ± 1. $\rho(X,Y)$ is a measure of the tendency for X and Y to be linearly related.

3.1. Some useful properties of expectations

For a single random variable X and any two constants a and b, it follows immediately from our definition of expectation that

$$E(a) = a , \tag{1}$$

$$E(aX + b) = aE(X) + b . \tag{2}$$

Equation (2) is a simple example of the general result that the expectation of the sum of a number of quantities (variables or constants) is the sum of the individual expectations. We may use this result to derive a useful expression for V(X).

$$
\begin{aligned}
V(X) &= E\{(X-\mu)^2\} \\
&= E(X^2 - 2\mu X + \mu^2) \\
&= E(X^2) - 2\mu^2 + \mu^2 \\
&= E(X^2) - \mu^2 & (3) \\
&= E(X^2) - \{E(X)\}^2 . & (4)
\end{aligned}
$$

It is often convenient to combine equation (3) with equation (3.13) to obtain an expression for V(X) in terms of the second derivative of the p.g.f., thus

$$V(X) = g''(1) + \mu - \mu^2, \text{ where } \mu = g'(1) .$$

Also

$$
\begin{aligned}
V(aX + b) &= E\{(aX - a\mu)^2\} \\
&= a^2 V(X) . & (5)
\end{aligned}
$$

For any two random variables X and Y, analogous arguments produce the following equations

$$E(X+Y) = E(X) + E(Y) \ , \tag{6}$$

$$\begin{aligned} C(X,Y) &= E\{(X-\mu_x)(Y-\mu_y)\} \\ &= E(XY - \mu_x Y - \mu_y X + \mu_x\mu_y) \\ &= E(XY) - 2\mu_x\mu_y + \mu_x\mu_y \\ &= E(XY) - \mu_x\mu_y \ . \end{aligned} \tag{7}$$

In the special case where X and Y are independent, E(XY) factorizes into E(X).E(Y). Hence for this special case, equation (7) shows that

$$C(X,Y) = 0 \ .$$

From equation (3.15), we have

$$\rho(X,Y) = 0 \ .$$

Note that the reverse is not necessarily true: the relation E(XY) = E(X).E(Y) and its corollary C(X,Y) = 0 do not imply that X and Y are independent. However, if X and Y are jointly normally distributed, then zero covariance does imply independence.

Consider the variance of the sum of two random variables

$$\begin{aligned} V(X+Y) &= E[\{(X+Y) - (\mu_x+\mu_y)\}^2] \\ &= E[\{(X-\mu_x) + (Y-\mu_y)\}^2] \\ &= E\{(X-\mu_x)^2\} + E\{(Y-\mu_y)^2\} + 2E\{(X-\mu_x)(Y-\mu_y)\} \\ &= V(X) + V(Y) + 2C(X,Y) \ . \end{aligned} \tag{8}$$

In the same way

$$V(X-Y) = V(X) + V(Y) - 2C(X,Y) \ . \tag{9}$$

If and only if C(X,Y) = 0, then

$$V(X+Y) = V(X) + V(Y) \tag{10}$$

and
$$V(X-Y) = V(X) + V(Y) . \tag{11}$$

Similar equations can be derived for any number of random variables.

<u>Example</u>. We illustrate the use of some of the equations in this section by evaluating the mean and variance of a rectangular distribution defined by the density

$$f(x) = \begin{cases} k \text{ for } a \leq x \leq b \\ 0 \text{ otherwise .} \end{cases}$$

Since
$$\int_{-\infty}^{+\infty} f(x)dx = 1 ,$$

$$k(b-a) = 1 ,$$

and so
$$k = 1/(b-a) .$$

The mean is

$$\mu = E(X) = \int_a^b \frac{xdx}{b-a}$$

$$= \frac{b+a}{2} ,$$

as we would expect, since the density is symmetrical about this value. Also

$$E(X^2) = \int_a^b \frac{x^2dx}{b-a}$$

$$= \frac{b^2+ab+a^2}{3} .$$

Therefore using equation (3.1.3), we obtain

$$V(X) = \frac{b^2+ab+a^2}{3} - \frac{(b+a)^2}{4}$$

$$= \frac{(b-a)^2}{12} .$$

In the particular case where $b = e$ and $a = -e$, we have

$$E(X) = 0 ,$$

$$V(X) = e^2/3 .$$

Thus, for the example given in Section 1, the variance of the rounding-error in measurements taken to the nearest mark is 1/12 since $e = 0.5$. Alternatively we may say that in this case

$$\sigma(X) = 0.29 .$$

4. Sampling distribution

An experiment is replicated n times producing independent, identically-distributed, observations $X_1, X_2, \ldots X_n$. We emphasize that the subscript simply labels the particular experiment in which the random variable is observed. We consider some function of these observations $T(X_1, X_2, \ldots X_n)$, which we shall write more briefly as $T(X)$. This random variable $T(X)$ possesses a probability distribution induced by, or as a consequence of, that of the X's. This distribution is called the sampling distribution of $T(X)$, or more simply of T.

Two such statistics of great importance are the sample mean, $\bar{X}$, and the sample variance, s^2, defined by the equations

$$\bar{X} = \Sigma X_i/n , \tag{1}$$

$$s^2 = s^2(X) = \Sigma(X_i - \bar{X})^2/(n-1) \tag{2}$$

$$= \{\Sigma X_i^2 - (\Sigma X_i)^2/n\}/(n-1) . \tag{3}$$

It is standard practice to use a small letter, s, for the sample variance although this departs from the standard notation of using a capital letter to represent a random variable. Thus, in a discussion of s^2, we have to rely on the context to make it clear whether it is being treated as a random variable or whether a particular realization is being referred to.

To illustrate these ideas, consider the results of 65

students who have each measured the enthalpy change, ΔH, accompanying the neutralization of NaOH using HCl. The data have been grouped to form 13 sets of 5 as shown in Table 3.1. The last two columns illustrate the sampling distribution of the two statistics, $\bar{X}$ and s^2.

Table 3.1. Grouped values of the enthalpy change accompanying the neutralization of NaOH using HCl.

Group	x_1	x_2	x_3	x_4	x_5	$\bar{x}$	s^2
1	56.9	59.2	56.3	58.0	56.9	57.46	1.32
2	53.8	55.4	58.0	59.6	55.5	56.46	5.34
3	58.4	55.0	55.7	56.6	57.2	56.58	1.74
4	58.0	56.4	57.6	57.5	55.0	56.90	1.48
5	57.7	58.5	58.9	57.8	57.4	58.06	0.38
6	54.8	56.4	55.2	60.3	57.1	56.76	4.76
7	57.1	60.4	58.9	55.5	54.7	57.32	5.55
8	58.6	57.8	58.0	55.5	55.6	57.10	2.09
9	58.9	59.8	60.0	57.1	56.4	58.44	2.61
10	59.5	57.7	60.0	57.6	56.8	58.32	1.86
11	57.2	58.2	57.4	55.7	59.1	57.52	1.60
12	55.4	56.1	57.7	56.9	59.2	57.06	2.17
13	55.1	56.8	55.7	61.6	58.3	57.50	6.74

$$x = -\Delta H/(\text{kJ mol}^{-1})$$

$$s^2 = s^2(\Delta H)/(\text{kJ}^2\ \text{mol}^{-2})$$

It is important to distinguish a <u>sample</u> parameter such as a mean or variance, from the corresponding <u>distribution</u> parameter. A distribution parameter, usually represented by a Greek letter, is a fixed quantity associated with the probability distribution. On the other hand, a sample parameter, usually represented by a Latin character, is a random variable, a function of the observations, varying from one sample to another. A sample parameter affords an <u>estimator</u> of the corresponding distribution parameter which may be unknown, but we shall consider this further when looking at estimation. Sometimes, when the context makes it clear which is meant, the adjective 'sample' or 'distribution' may be dropped.

Even without knowing the distribution of the X's, we can say something about certain parameters in the distributions of $\bar{X}$ and s^2. Thus

$$E(\bar{X}) = \Sigma\mu/n = \mu = E(X) , \quad (4)$$

Remembering that the X's are independent and using equations (3.1.5) and (3.1.10), we obtain

$$V(\bar{X}) = \Sigma\sigma^2/n^2 = n\sigma^2/n^2 = \frac{\sigma^2}{n} . \quad (5)$$

Hence, as n increases, the distribution of $\bar{X}$ becomes more concentrated about the mean value , μ , but this occurs slowly; increasing n fourfold, only halves the standard deviation of $\bar{X}$. To derive the expectation of s^2, we make use of the identity

$$\Sigma(X_i-a)^2 = \Sigma[\{(X_i-\bar{X}) + (\bar{X}-a)\}^2] = \Sigma(X_i-\bar{X})^2 + n(\bar{X}-a)^2 , \quad (6)$$

where a is any constant. If we put $a = \mu$ and then take expectations of both sides of the resulting equation, we obtain

$$n\sigma^2 = E\{(n-1)s^2\} + nV(\bar{X})$$

$$= (n-1)E(s^2) + \sigma^2 .$$

Therefore

$$E(s^2) = \sigma^2 . \tag{7}$$

Just as we call s^2, the sample variance, so s is called the <u>sample standard deviation</u> and both s^2 and s are said to have n-1 <u>degrees of freedom</u>.

In Chapter 4, Section 10 we shall consider the distribution of $\bar{X}$ and s^2 when the X's are normally distributed.

CHAPTER 4

SOME IMPORTANT PROBABILITY DISTRIBUTIONS

In this chapter, we shall deal with a number of probability distributions which are important in probability and statistics. As in the previous chapter, we shall be concerned primarily with mathematics, stating formulae and deriving properties for use later in our analysis of data. Several of these probability distributions will already be familiar to our readers.

1. The binomial distribution

Suppose we divide the possible outcomes of an experiment into two complementary events, say A and $\bar{A}$; for example, we may have 'heads' and 'tails' when tossing a coin, or {1,2} and {3,4,5 or 6} when tossing a die, or 'less than or equal to 3.5 cm' and 'greater than 3.5 cm' when measuring a length. We shall call an occurrence of the event A, a <u>success</u> (S), and of $\bar{A}$, a <u>failure</u> (F). We shall represent the probability of a success in a single trial by p, and that of a failure by $q = 1 - p$.

Suppose now that we perform n such independent experiments, or n binomial <u>trials</u> as they are called. The total number of successes, K say, out of n such trials is then a discrete random variable on the sample space, {0,1,...n}. We say that K has the binomial distribution B(p,n), since, as we shall see, the probabilities $P(K = k)$, $k = 0,1,\ldots n$, are found from the binomial expansion of $(p + q)^n$.

The probability of any particular sequence, say SSFSSSFFSF..., comprised of k S's and n - k F's is $ppqpppqqpq\ldots = p^k q^{n-k}$, because the trials are independent. The number of sequences containing just k S's is the number of ways of choosing k items from n, ${}^nC_k$. Thus, following the arguments of Chapter 3, to obtain $P(K = k)$, we have to sum the probabilities of the ${}^nC_k$ simple events which go to make up the event $K = k$. Therefore

$$P(K = k) = P(k) = {}^{n}C_{k}\, p^{k}\, q^{n-k} = {}^{n}C_{k}\, p^{k}(1-p)^{n-k}, \quad k = 0,1,\ldots n, \tag{1}$$

where ${}^{n}C_{k} = n!/\{k!(n-k)!\}$, so that P(k) is the term in p^k in the expansion of $(p + q)^n$.

The total probability for all k is

$$\sum_k P(k) = \sum_k {}^{n}C_{k}\, p^{k}\, q^{n-k} = (p + q)^n = 1^n = 1 .$$

It is sometimes convenient to calculate the probabilities of successive values of k using a recurrence relationship. Dividing P(k) by P(k-1) and cancelling out common factors gives

$$\frac{P(k)}{P(k-1)} = \frac{(n - k + 1)}{k} \cdot \frac{p}{q}, \text{ and } P(0) = q^n .$$

These relations may be used to calculate P(0), P(1), P(2), etc. in succession. When using this method it is advisable to check the last of the sequence either by independently calculating it or by checking that the sum of the probabilities for all k is one, apart from rounding-off error.

Example 1. Suppose that the probability that any single performance of a particular experiment will yield a usable result is 0.6. If the experiment is performed 5 times, what is the probability distribution of the number of usable results? What is the probability of at least 2 usable results?

First we calculate the individual probabilities as below from which we construct Figure 4.1:

	direct method			recurrence method ($p/q = 0.6/0.4 = 3/2$)
P(0) =	$1 \times 0.6^0 \times 0.4^5$	= 0.01024		
P(1) =	$5 \times 0.6^1 \times 0.4^4$	= 0.07680	←	P(0) x 5/1 x 3/2
P(2) =	$10 \times 0.6^2 \times 0.4^3$	= 0.23040	←	P(1) x 4/2 x 3/2
P(3) =	$10 \times 0.6^3 \times 0.4^2$	= 0.34560	←	P(2) x 3/3 x 3/2
P(4) =	$5 \times 0.6^4 \times 0.4^1$	= 0.25920	←	P(3) x 2/4 x 3/2
P(5) =	$1 \times 0.6^5 \times 0.4^0$	= 0.07776	←	P(4) x 1/5 x 3/2

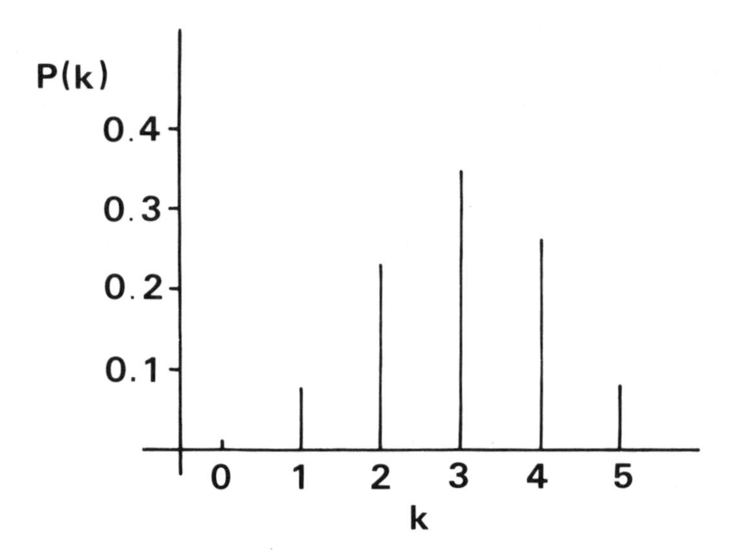

Figure 4.1. Probability of k successes in 5 binomial trials with p = 0.6

To obtain $P(K \geq 2)$, we write

$$\begin{aligned} P(K \geq 2) &= P(2) + P(3) + P(4) + P(5) \text{ , or more simply} \\ &= 1 - P(0) - P(1) \\ &= 0.91296 \end{aligned}$$

Example 2. We draw attention to the application of the binomial distribution to sampling with replacement. We recall the problem in Chapter 2, Section 3 of drawing two discs from a bag containing 3 red and 4 blue discs, but this time we replace a disc after it has been withdrawn and its colour noted. If we label the withdrawal of a red disc as a success, then our problem is simply that of evaluating P(1) for the distribution B(3/7,2). This is ${}^2C_1(3/7)^1(4/7)^1 = 24/49$.

1.1. Mean and Variance

We shall derive these two quantities in three different ways, since the methods themselves are of interest.

(i) Direct method

$$\mu = E(K) = \sum_{k=0}^{n} k\ P(k)$$

Thus,

$$\mu = \sum_{k=1}^{n} k\, {}^{n}C_{k}\, p^{k}\, q^{n-k}$$

$$= \sum_{k=1}^{n} \frac{kn!}{k!(n-k)!}\, p^{k}\, q^{n-k}$$

$$= np \sum_{k-1=0}^{n-1} {}^{n-1}C_{k-1}\, p^{k-1}\, q^{n-1-(k-1)}$$

$$= np(p + q)^{n-1}$$

$$= np\ .$$

For the variance, we evaluate $E\{K(K-1)\}$ so as to find $E(K^2)$.

$$E\{K(K-1)\} = \sum_{k=2}^{n} k(k-1)\, P(k)$$

$$= \sum_{k=2}^{n} k(k-1)\, {}^{n}C_{k}\, p^{k}\, q^{n-k}$$

$$= n(n-1)p^{2} \sum_{k-2=0}^{n-2} {}^{n-2}C_{k-2}\, p^{k-2}\, q^{n-2-(k-2)}$$

$$= n(n-1)p^{2}\ .$$

Now

$$E(K^2) = E\{K(K-1)\} + E(K)$$

$$= n(n-1)p^{2} + np\ .$$

Also

$$\sigma^{2} = E(K^{2}) - \{E(K)\}^{2}$$

$$= n(n-1)p^{2} + np - n^{2}p^{2}$$

$$= np(1-p)$$

$$= npq\ .$$

(ii) Using the p.g.f.

From Chapter 3, we have

$$g(t) = E(t^K) = \sum_k {}^nC_k\, p^k q^{n-k} t^k$$

$$= (pt + q)^n$$

$$g'(t) = np(pt + q)^{n-1}$$

$$g''(t) = n(n-1)\, p^2(pt + q)^{n-2} .$$

Hence

$$g'(1) = np = E(K) = \mu, \text{ and}$$

$$g''(1) = n(n-1)p^2 = E\{K(K-1)\},$$

which leads to $\sigma^2 = npq$ as shown in (i).

(iii) Using an indicator variable

Let I_r = number of S's from the rth trial.

$$= \begin{cases} 1 \text{ with probability } p \\ 0 \text{ with probability } q . \end{cases}$$

Now

$$E(I_r^a) = 1^a p + 0^a q$$

$$= p, \text{ for any } a > 0 .$$

Hence

$$E(I_r) = p ,$$

and

$$V(I_r) = E(I_r^2) - \{E(I_r)\}^2 ,$$

$$= p - p^2 = pq .$$

But

$$K = \sum_{r=1}^{n} I_r ,$$

so that

$$E(K) = \mu = n\,E(I_r) = np ,$$

and

$$V(K) = \sigma^2 = n\,V(I_r) = npq .$$

1.2. Additivity property

If K_1 and K_2 are independently distributed as $B(p,n_1)$ and $B(p,n_2)$, then their p.g.f's are $(pt + q)^{n_1}$ and $(pt + q)^{n_2}$. Hence the p.g.f. of K_1+K_2 is $(pt + q)^{n_1+n_2}$, so that, by the uniqueness theorem of Chapter 3, K_1+K_2 is distributed as $B(p,n_1+n_2)$. A similar result holds for the sum of m independent, binomially distributed random variables.

1.3. Relationships to other distributions

The binomial distribution can be used to approximate to the hypergeometric distribution (see Chapter 2, Section 3). When the sample size, s, is small compared to the batch size N, then the number of 'specials' sampled, D, is approximately distributed as $B(r/N,s)$, where r is the number of specials in the batch. We may say that sampling without replacement (when the hypergeometric distribution applies) is well approximated by sampling with replacement (when the binomial applies) if s/N is small; this makes sense physically since then the proportion of specials in the batch remains almost constant during the sampling.

We shall discuss the relationship of the binomial to the Poisson and normal distributions in the appropriate sections.

2. The Poisson distribution

This discrete distribution relates to the number of events that occur per given segment of time or space when the events occur randomly in time or space at a certain average rate. Some examples of types of random variables for which the Poisson distribution is often assumed to apply are: the number of particles emitted from a radioactive source in a given time, the number of faults per given length of yarn, the number of typing errors per page of manuscript, the number of vehicles passing a given point on a road in a given time in light traffic, the number of goals scored by a particular team in a football match, the number of calls received at a telephone exchange in a given time period. The sample space for the type of random variable under discussion thus consists of the integers $\{0,1,2,...\}$.

As in Section 1, we shall use K to represent the random variable on this sample space. Then, we define the Poisson distribution as that distribution in which the probability that $K = k$ is given by

$$P(K = k) = P(k) = m^k e^{-m}/k! , \quad k = 0,1,2,... \tag{1}$$

We shall refer to such a situation by writing $K \sim Pn(m)$, i.e. K is distributed according to the Poisson distribution with parameter m. We emphasize that P(k) gives the probability that k events occur in the chosen segment of time or space. Clearly

$$\sum_0^\infty P(k) = e^{-m} \sum_0^\infty m^k/k! = e^{-m} e^m = 1 .$$

The recurrence relation for the calculation of the P(k)'s is

$$P(k) = P(k-1).m/k \tag{2}$$

$$P(0) = e^{-m} . \tag{3}$$

It can be shown that, if we increase the size of each segment by a factor a, the number of events per segment is distributed as Pn(am).

2.1. Mean and variance

(i) Direct method

$$\mu = E(K) = \sum_{k=0}^\infty k\, m^k e^{-m}/k!$$

Thus,

$$\mu = \sum_{k=1}^{\infty} m^k \, e^{-m}/(k-1)!$$

$$= m \, e^{-m} \sum_{k-1=0}^{\infty} m^{k-1}/(k-1)!$$

$$= m \, e^{-m} \, e^{m}$$

$$= m \; .$$

We observe that the parameter m is the mean number of events which occur per given segment of time or space.

To calculate the variance, we first evaluate $E\{K(K-1)\}$.

$$E\{K(K-1)\} = \sum_{k=0}^{\infty} k(k-1) \, m^k \, e^{-m}/k!$$

$$= \sum_{k=2}^{\infty} m^k \, e^{-m}/(k-2)!$$

$$= m^2 \, e^{-m} \sum_{k-2=0}^{\infty} m^{k-2}/(k-2)!$$

$$= m^2 \; .$$

But,

$$\sigma^2 = E\{K(K-1)\} + E(K) - \{E(K)\}^2$$

$$= m^2 + m - m^2$$

$$= m \; .$$

Thus the mean and variance of a Poisson distribution are the same.

(ii) Using the p.g.f.

$$g(t) = E(t^K) = \sum_0^{\infty} (mt)^k e^{-m}/k!$$

$$= e^{mt} e^{-m}$$

$$= e^{m(t-1)} .$$

$$g'(t) = m e^{m(t-1)} ,$$

$$g''(t) = m^2 e^{m(t-1)} .$$

Hence

$$g'(1) = m = \mu ,$$

and

$$g''(1) = m^2 = E\{K(K-1)\},$$

which leads to $\sigma^2 = m$ as shown in (i).

2.2. Additivity property

If K_1 and K_2 are independently distributed as $Pn(m_1)$ and $Pn(m_2)$, the p.g.f. of K_1+K_2 is

$$\exp\{m_1(t-1)\}\exp\{m_2(t-1)\} = \exp\{(m_1+m_2)(t-1)\} .$$

Hence K_1+K_2 is distributed as $Pn(m_1+m_2)$. An analogous result holds for the sum of any number of independent Poissonly distributed random variables.

2.3. Relationship to other distributions

The Poisson distribution affords a useful approximation to the binomial $B(p,n)$ when p is small and n is large. Then the number of successes is approximately distributed as $Pn(np)$.

For the normal approximation to the Poisson distribution, see Section 6.5.

2.4. Example. A laboratory counter was arranged to measure the cosmic ray 'background'. For the purpose of this example, it was set up to record the number of particles arriving in intervals of 0.1 seconds. A very large number of measurements were made so that we could reasonably take the mean of the measurements, viz. 11.60, as the parameter m of the distribution. Figure 4.2. shows a plot of P(k) against k for this particular Poisson distribution. We observe from this figure that the distribution is not quite symmetrical. If m were smaller, the distribution would be more skew.

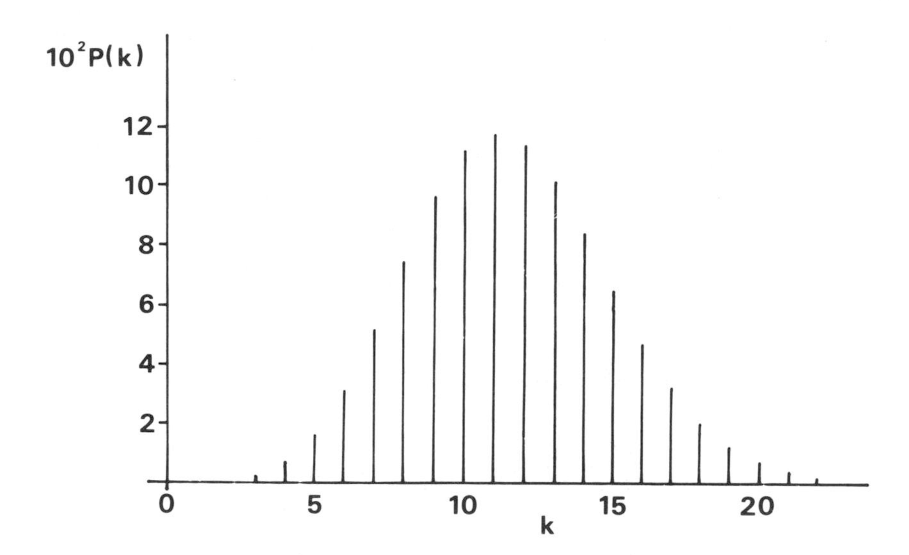

Figure 4.2. Probability of k events for the distribution Pn(11.6)

When the experiment was repeated with a radioactive source close to the detector, the mean of the number of particles arriving in the same interval of time taken over many measurements was 98.73. Now the numbers of particles arriving at the detector from the radioactive source and from outer space are independent. Hence by the additivity theorem, we deduce that the number of particles arriving from the radioactive source alone is distributed as Pn(87.13). It is important that the two random variables added together must be independent for the additivity theorem to apply.

3. The Poisson process

The Poisson process is a process in which events occur randomly in time or space; usually, we shall think in terms of time in this section. We shall see that the numbers of events per given time have a Poisson distribution while the intervals between consecutive events have an exponential distribution.

We assume that the probability of the occurrence of an event in the time interval $(t, t+\delta t)$ is $\lambda\delta t + o(\delta t)$, where λ is simply a constant characteristic of the process. Here δt is small and $o(\delta t)$ means 'small compared with δt'. We also assume that the probability of the occurrence of more than one event in the interval in question is $o(\delta t)$. Note that the probability of an event in $(t, t+\delta t)$ is independent of t and of what has occurred up to t.

Suppose we consider the probability of the occurrence of n events in the interval $(0, t+\delta t)$ where $n \geq 1$. Under the above assumptions, we need only consider the probabilities of two alternative ways of reaching this situation,

A: n events occur in the interval $(0,t)$ and none in the next δt,

B: $n-1$ events occur in the interval $(0,t)$ and one in the next δt,

since we do not need to consider other possibilities of very small probability. Let us write $P(n,t)$ for the probability that n events have occurred in the interval $(0,t)$. We have

$$P(A) = P(n,t).(1 - \lambda\delta t) + o(\delta t),$$

$$P(B) = P(n-1,t).\lambda\delta t + o(\delta t),$$

also

$$P(n,t+\delta t) = P(A) + P(B) + o(\delta t) .$$

Hence

$$P(n,t+\delta t) = P(n,t).(1 - \lambda\delta t) + P(n-1,t).\lambda\delta t + o(\delta t), \quad (1)$$

which gives

$$\frac{P(n,t+\delta t) - P(n,t)}{\delta t} = \lambda\{P(n-1,t) - P(n,t)\} + \frac{o(\delta t)}{\delta t}$$

In the limit, as $\delta t \to 0$,

$$\dot{P}(n,t) = \lambda\{P(n-1,t) - P(n,t)\} . \qquad (2)$$

Equation (2) may be recast, thus

$$e^{-\lambda t}.\frac{d}{dt}\left\{e^{\lambda t} P(n,t)\right\} = \lambda P(n-1,t),$$

which on integration between the limits 0 (where $P(0,0) = 1$) and t yields

$$e^{\lambda t} P(n,t) = \lambda\int_0^t e^{\lambda t} P(n-1,t)\, dt . \qquad (3)$$

Equation (3) is a recurrence formula by which we may obtain successively $P(1,t)$, $P(2,t)$,... $P(n,t)$, given $P(0,t)$. For the case $n = 0$, we have

$$P(0,t+\delta t) = P(0,t).(1 - \lambda\delta t) + o(\delta t) .$$

Hence

$$-\lambda = \frac{\dot{P}(0,t)}{P(0,t)} = \frac{d}{dt}\{\ln P(0,t)\} .$$

Therefore

$$P(0,t) = e^{-\lambda t} . \qquad (4)$$

Using equations (3) and (4), it can easily be demonstrated by induction that

$$P(n,t) = \frac{(\lambda t)^n e^{-\lambda t}}{n!} . \qquad (5)$$

Hence the number of occurrences in the time interval $(0,t)$ is distributed as $Pn(\lambda t)$ as stated in the introduction to this section.

We now consider the distribution of the time to the first occurrence of an event, using equation (4). The probability that the first event will occur in the interval $(t,t+\delta t)$ is the probability that none have occurred up to time t multiplied by the probability that one event occurs in the interval (again, neglecting alternatives of very small probability);

that is

$$P(O,t).\lambda\delta t + o(\delta t) .$$

Substituting for P(O,t) from equation (4) gives for the probability in question

$$\lambda e^{-\lambda t} \delta t + o(\delta t) .$$

Hence, the density of the distribution of the time to the first occurrence of an event, which we shall represent by $f_1(t)$, is given by

$$f_1(t) = \lambda e^{-\lambda t} , \tag{7}$$

the density of an exponential distribution with mean λ^{-1} (see Section 4). Here, the starting time may be just after an event has occurred or any other time.

It is easy to extend this discussion to the distribution of the time elapsed to the nth occurrence of an event, once again taking the initial time at any point. The probability that the nth event occurs in the interval $(t,t+\delta t)$ is the probability that (n-1) events occur in (O,t) and one occurs in $(t,t+\delta t)$, again neglecting other possibilities of very small probability. Now this is P(B) viz.

$$P(n-1,t).\lambda\delta t = \frac{(\lambda t)^{n-1} e^{-\lambda t} \lambda\delta t}{(n-1)!} ,$$

using equation (5). Hence the density of the distribution of the time to the nth event, $f_n(t)$, is

$$f_n(t) = (\lambda t)^{n-1} e^{-\lambda t} \lambda/(n-1)! \tag{8}$$

We shall see later (Section 5) that this is the density of the sum of n independent exponential variates, each with mean λ^{-1}.

4. The exponential distribution

This is the distribution of the time elapsed, space covered, etc. before a randomly located event occurs. Thus the time elapsed between consecutive events in a Poisson process has an exponential distribution, so examples could be listed corresponding to those given for the Poisson distribution. Often, the lifetime of a component in a piece of apparatus is assumed to have such a distribution. Another example of an exponentially distributed random variable is the distance travelled between successive collisions in a low pressure gas.

It will be clear from these examples that here we have a continuous random variable for which the sample space is the positive half of the real line, $x \geq 0$. We say that our random variable X has the exponential distribution if its density $f(x)$ is given by

$$f(x) = a\, e^{-ax}, \qquad x \geq 0,\ a > 0 . \tag{1}$$

Obviously

$$\int_0^\infty f(x)\, dx = 1,$$

as is required for a density. Clearly $F(x) = 1 - e^{-ax}$.

4.1. Mean and variance

$$\begin{aligned} \mu = E(X) &= \int_0^\infty ax\, e^{-ax}\, dx \\ &= \left[x\, e^{-ax} + \int e^{-ax}\, dx\right]_0^\infty \\ &= 1/a . \end{aligned}$$

To find the variance, we first evaluate $E(X^2)$.

$$E(X^2) = \int_0^\infty a\, x^2\, e^{-ax}\, dx .$$

Thus

$$E(X^2) = \left[x^2 e^{-ax} + \int 2x\, e^{-ax}\, dx \right]_0^\infty$$

$$= 2/a^2 .$$

Hence

$$\sigma^2 = 2/a^2 - 1/a^2$$

$$= 1/a^2 .$$

4.2. Relationship to other distributions

We have already seen how the exponential distribution is connected with the Poisson process. It is also closely related to the Gamma distribution, of which it is the simplest case.

4.3. Example. Ditertiary butyl peroxide (DTBP) decomposes at 154.6°C in the gas phase by a first order process with rate constant $k=3.46 \times 10^{-4} s^{-1}$. By this, we mean that the number of molecules, $N(t)$, of DTBP remaining at time t after the reaction has started is given by

$$N(t) = N(0)\exp(-kt) ,$$

where $N(0)$ is the number present at $t = 0$. The decrease in the number of molecules of DTBP, $-dN(t)$, in the interval $(t,t+dt)$ is thus

$$-dN(t) = N(0)\, k \exp(-kt)\, dt . \tag{1}$$

But $-dN(t)$ is the number of molecules out of our original total which survive for a time T and no longer, where $t < T \leq t+dt$. Therefore, the probability that one of our original molecules will survive to this time, viz,

$$P(t < T \leq t+dt) = -\frac{dN(t)}{N(0)}$$

$$= k \exp(-kt)\, dt ,$$

from equation (1).

Hence, the density of the survival time is

$$f(t) = k \exp(-kt) ,$$

which is plotted in Figure 4.3 for this particular case. We note that the average survival time of the DTBP molecules is $1/k = 10^4 s/3.46 = 2.89 \times 10^3$ s; we also note that 50% of the molecules have disappeared in 2.00×10^3 s, a time which is usually called the half-life, the time satisfying $F(t) = 0.5$, that is $t_{\frac{1}{2}} = (\ln 2)/k$. Thus less than half the molecules survive for the average time, the precise fraction being, of course, $1/e \simeq 0.37$.

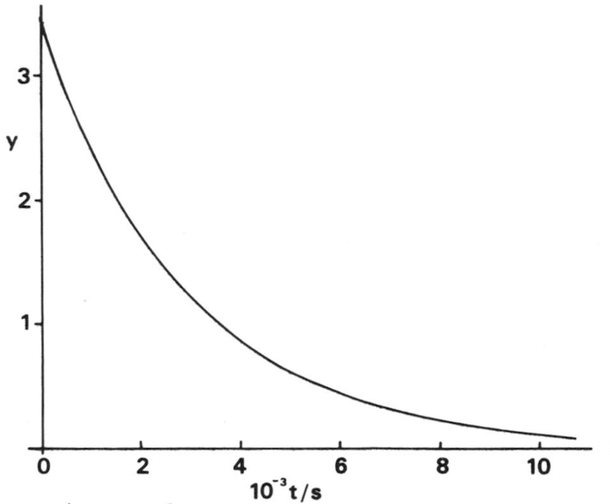

Figure 4.3. $y = 10^4 f(t)/s^{-1}$, the density for the survival time of DTBP molecules at 154.6°C, plotted against $10^{-3}t/s$

5. The gamma distribution

This distribution is related to the exponential distribution and is exemplified in the distribution of the sum of n independent exponential variates each with the same mean - see equation (3.8). This distribution is also related to the chi-squared distribution (see Section 7) and could arise in other ways. We say that X has the gamma distribution if

$$f(x) = a^b x^{b-1} e^{-ax}/\Gamma(b), \quad x \geq 0; \; a, b > 0 . \tag{1}$$

We shall denote this by $X \sim Gm(a,b)$. Here b, which is often an integer, is called the number of degrees of freedom of X or of the distribution of X. The ratio $\Gamma(b+1)/\Gamma(b) = b$, so that $\Gamma(b+2)/\Gamma(b) = (b+1)b$, etc. If b is an integer then $\Gamma(b) = (b-1)!$ We thus see that the gamma distribution with one degree of freedom is the same as the exponential distribution.

5.1. Mean and variance

We shall use

$$\int_0^\infty x^r e^{-x} dx = \Gamma(r+1) .$$

Now

$$\begin{aligned} E(X^s) &= \int_0^\infty a^b x^{b+s-1} e^{-ax} dx/\Gamma(b) \\ &= a^{-s} \int_0^\infty u^{b+s-1} e^{-u} du/\Gamma(b) \\ &= a^{-s} \Gamma(b+s)/\Gamma(b) . \end{aligned}$$

Therefore

$$\begin{aligned} \mu &= E(X) = \frac{\Gamma(b+1)}{a\Gamma(b)} \\ &= b/a , \end{aligned}$$

and

$$\begin{aligned} \sigma^2 &= E(X^2) - \{E(X)\}^2 \\ &= \frac{\Gamma(b+2)}{a^2\,\Gamma(b)} - \frac{1}{a^2}\left\{\frac{\Gamma(b+1)}{\Gamma(b)}\right\}^2 \\ &= b(b+1)/a^2 - b^2/a^2 \\ &= b/a^2 . \end{aligned}$$

5.2. Additivity property

First, we derive an expression for the m.g.f. of the Gm(a,b) distribution,

$$\begin{aligned}
\text{m.g.f.} &= E\left(e^{tX}\right) \\
&= \int_0^\infty a^b\, x^{b-1}\, e^{-ax+tx}\, dx/\Gamma(a) \\
&= (a-t)^{-b}\, a^b \int_0^\infty u^{b-1}\, e^{-u}\, du/\Gamma(a) \\
&= (1 - t/a)^{-b} .
\end{aligned}$$

If we have two random variables X_1 and X_2 independently distributed as $Gm(a,b_1)$ and $Gm(a,b_2)$, then X_1+X_2 has m.g.f., $(1 - t/a)^{-b_1-b_2}$. Hence $X_1+X_2 \sim Gm(a,b_1+b_2)$. Similarly, if $\{Y_i\}$ are n independent $Gm(a,1)$ variates, $\Sigma Y_i \sim Gm(a,n)$ as already stated.

5.3. A connection between the gamma and Poisson distributions

We consider the random variable $Z \sim Gm(1,m)$, where m is an integer. Then

$$\begin{aligned}
P(Z > c) &= \int_c^\infty z^{m-1} e^{-z} dz/(m-1)!, \quad \text{for any } c > 0 , \\
&= \left[-z^{m-1}e^{-z} + \int (m-1)\, z^{m-2}e^{-z}dz\right]_c^\infty /(m-1)! \\
&= \Big[-z^{m-1}e^{-z} - (m-1)\, z^{m-2}e^{-z} \\
&\qquad + \int (m-1)(m-2)\, z^{m-3}e^{-z}dz\Big]_o^\infty /(m-1)! \\
&= \text{etc.} \\
&= - \left[\{z^{m-1}/(m-1)! + z^{m-2}/(m-2)! + \ldots + z + 1\}\, e^{-z}\right]_c^\infty \\
&= e^{-c}\{1 + c + c^2/2! + \ldots + c^{m-1}/(m-1)!\} \qquad (1) \\
&= P(K < m-1), \text{ where } K \sim Pn(c) .
\end{aligned}$$

We could have arrived at this result by considering a Poisson process in which events occur at an average rate of 1 per second. In this case, Z seconds would represent the waiting time until the occurrence of the mth event. The probability that this waiting time is greater than c seconds is just the probability that not more than m-1 events have occurred in the time interval (0, c seconds), that is

$$P(Z > c) \quad = \quad P(K \leq m-1), \quad \text{where } K \sim Pn(c) \ ,$$

as derived formally earlier.

Now if $Y \sim Gm(a,m)$, then $Z = aY$ is distributed as $Gm(1,m)$, a result which applies even if m is not an integer. Hence

$$P(Y > c) \quad = \quad P(Z > ac) \ ,$$

which is given by equation (1) replacing c by ac.

5.4. Example. A car is fifth in a queue of vehicles waiting at a toll booth. Its waiting time is the sum of the four service times for the preceding vehicles. The service times are independently exponentially distributed with mean 20 seconds. What is the probability that the car in question will have to wait more than 90 seconds?

If the service time is T seconds, we know from the information given, that T is distributed as $a\exp(-at)$ where $E(T) = 1/a = 20$. Hence $a = 1/20$. Thus if the waiting time is W seconds, W is the sum of 4 independent exponential variates each with parameter 1/20. Hence $W \sim Gm(1/20,4)$ and $P(W > 90)$ is obtained by putting $c = 90/20$ in equation (1):

$$\begin{aligned} P(W > 90) &= e^{-4.5}(1 + 4.5 + 4.5^2/2 + 4.5^3/6) \\ &= 0.01111 \times 30.8125 \\ &= 0.3423 \ . \end{aligned}$$

6. The normal distribution

This is the most important and widely-used distribution in statistics, both directly for the distribution of continuous random variables, and as an approximation to the distributions of observations and derived statistics (continuous and discrete).

We say that X is normally distributed with mean μ and variance σ^2 if it has the probability density

$$f(x) = (2\pi\sigma^2)^{-\frac{1}{2}} \exp\{-(x - \mu)^2/(2\sigma^2)\}, \quad -\infty < x < \infty . \tag{1}$$

We shall often write $X \sim N(\mu,\sigma^2)$ for such a random variable.

It is convenient to consider the random variable Z, which we term the 'standardised form' of X, defined by

$$Z = \frac{X - \mu}{\sigma} .$$

We obtain the density of Z, $\phi(z)$, by using the fact that

$$P(z < Z \leq z+dz) = P(x < X \leq x+dx) .$$

It follows that

$$\phi(z) = f(x)/\left|\frac{dz}{dx}\right| . \tag{2}$$

Here the modulus of the derivative is taken since a probability density can never be negative. Such a relation applies whenever a density is obtained from that of a functionally-related random variable; apart from the use of the modulus, the procedure is the same as that used when transforming a variable in an integration.* We complete the derivation of

---

* If Z(X) = Z, for a given Z, can be satisfied by more than one value of X, then the right-hand side of equation (2) should be summed over these roots. For example, if $Z = X^2$, the density of Z is

$$f(x)/|2x| + f(-x)/|-2x|$$

the density of Z by writing the right-hand side of equation (2) in terms of z, thus

$$\phi(z) \quad = \quad (2\pi)^{-\frac{1}{2}} \exp(-z^2/2), \qquad -\infty < z < \infty . \tag{3}$$

Z is often termed the <u>standard normal variate</u>. Comparing equations (1) and (3), we may write $Z \sim N(0,1)$. Figure 4.4 shows the shape of the density curve for Z. We observe that it

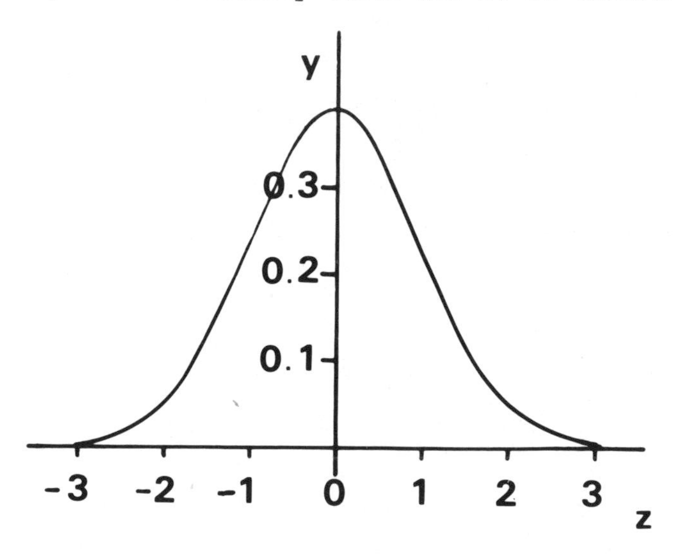

Figure 4.4. y = f(z), the standard normal density function

is symmetric about z = 0 and bell-shaped, and that it almost coincides with the axis outside z = ± 3. We may easily show that the area under this curve, viz.

$$\int_{-\infty}^{\infty} \phi(z)dz \ ,$$

is unity. For example, we may square the integral to obtain

$$(2\pi)^{-1} \int_{-\infty}^{\infty}\int_{-\infty}^{\infty} e^{-\frac{1}{2}(u^2 + v^2)} dudv \ ,$$

then transform the result to polar coordinates, and finally perform the integration.

6.1. Mean and variance

We have for Z

$$E(Z) = \int_{-\infty}^{\infty} (2\pi)^{-\frac{1}{2}} z e^{-\frac{1}{2}z^2} dz \qquad (1)$$

$$= 0, \text{ because the integrand is odd.}$$

Alternatively we note that equation (1) may be rewritten

$$E(Z) = -\int_{z=-\infty}^{z=\infty} d\phi(z)$$

$$= \Big[\phi(z)\Big]_{z=-\infty}^{z=\infty}$$

$$= 0 .$$

The variance may be found as in previous sections by first evaluating $E(Z^2)$. We have

$$E(Z^2) = \int_{-\infty}^{\infty} (2\pi)^{-\frac{1}{2}} z^2 e^{-\frac{1}{2}z^2} dz$$

$$= -\int_{z=-\infty}^{z=\infty} z d\phi(z)$$

$$= -\Big[z\phi(z) - \int \phi(z) dz\Big]_{z=-\infty}^{z=\infty}$$

$$= 1 .$$

Hence

$$V(Z) = E(Z^2) - \{E(Z)\}^2$$

$$= 1 . \qquad (2)$$

These two results provide a quick way of finding E(X) and V(X). We write

$$X = \sigma Z + \mu .$$

Hence

$$\left.\begin{aligned} E(X) &= \sigma E(Z) + \mu = \mu \\ \text{and} \quad V(X) &= \sigma^2 V(Z) = \sigma^2 \, . \end{aligned}\right\} \qquad (3)$$

This confirms our earlier statement that the parameters μ and σ^2 appearing in the expression $X \sim N(\mu,\sigma^2)$ are the mean and variance of the distribution.

6.2. The m.g.f. of Z and X

The m.g.f. of Z is

$$E\left(e^{tZ}\right) = \int_{-\infty}^{\infty} (2\pi)^{-\frac{1}{2}} e^{-\frac{1}{2}z^2+tz} dz$$

$$= e^{\frac{1}{2}t^2} \int_{-\infty}^{\infty} (2\pi)^{-\frac{1}{2}} e^{-\frac{1}{2}(z-t)^2} dz$$

$$= e^{\frac{1}{2}t^2} \, .$$

Hence, using Chapter 3, Section 3, the m.g.f. of X is

$$\exp(\mu t + \sigma^2 t^2/2) \, .$$

6.3. Additivity property

If X_1 and X_2 are independently distributed as $N(\mu_1,\sigma_1^2)$ and $N(\mu_2,\sigma_2^2)$, then the m.g.f. of X_1+X_2 is

$$\exp\{(\mu_1+\mu_2)t + (\sigma_1^2+\sigma_2^2)t^2/2\} \, .$$

Hence, the distribution of X_1+X_2 is $N(\mu_1+\mu_2,\sigma_1^2+\sigma_2^2)$.

More generally, if $X_1,X_2,\ldots X_n$ are independently distributed with $X_i \sim N(\mu_i,\sigma_i^2)$ for $i = 1,2,\ldots n$, then $\Sigma a_i X_i$ has m.g.f. $\exp\{t\Sigma a_i\mu_i + t^2\Sigma a_i^2\sigma_i^2/2\}$. Hence the distribution of the linear combination $\Sigma a_i X_i$ is $N(\Sigma a_i\mu_i,\Sigma a_i^2\sigma_i^2)$. We can be more general still, for a similar result holds if the X's are not independent; then the linear combination is distributed as

$$N\left\{\sum_i a_i\mu_i,\ \sum_{ij}\sum a_i a_j C(X_i,X_j)\right\} ,$$

which we may again prove by using the m.g.f.

6.4. The distribution function of Z, $\Phi(z)$

The standard normal integral

$$\Phi(z) = \int_{-\infty}^{z} \phi(y)\,dy .$$

This distribution function is widely tabulated, usually for positive z values only, because of its symmetry. The relation

$$\Phi(-z) = 1 - \Phi(z) \tag{1}$$

serves to obtain values of the distribution function for negative z values. A notation which we shall frequently employ is $z(1-\alpha)$, meaning the $1-\alpha$ quantile of the distribution of Z. Thus

$$\Phi\{z(1-\alpha)\} = 1 - \alpha .$$

Now, for any interval (a,b) such that $a < b$,

$$P(a < X \leq b) = P\{(a-\mu)/\sigma < Z \leq (b-\mu)/\sigma\}$$

$$= \Phi\{(b-\mu)/\sigma\} - \Phi\{(a-\mu)/\sigma\} .$$

Hence, the probability of X lying in any given interval can be obtained easily when μ and σ are known, using a standard normal integral table.

Examples of the use of tables of $\Phi(z)$

(i) A certain physical quantity, X, is distributed as N(3,4). What is the probability of observing (a) $X > 3.5$ (b) $X < 1.2$ (c) $2.5 < X < 3.5$?

We put $Z = (X-3)/2$ so that $Z \sim N(0,1)$. We proceed as follows.

(a) $P(X > 3.5) = P(Z > 0.25) = 1 - \Phi(0.25)$
$= 1 - 0.5987$
$= 0.4013.$

(b) $$P(X < 1.2) = P(Z < -0.9) = \Phi(-0.9) = 1 - \Phi(0.9) = 1 - 0.8159 = 0.1841.$$

(c) $$P(2.5 < X < 3.5) = P(-0.25 < Z < 0.25) = \Phi(0.25) - \Phi(-0.25) = 2\Phi(0.25) - 1 = 1.1974 - 1 = 0.1974.$$

The three probabilities are thus (a) 0.4013, (b) 0.1841, (c) 0.1974.

(ii) 6.3 percent of observations from a normal distribution have values above 3.287 and 51.2 percent have values above 2.897. What are the mean and variance of the distribution?

Suppose $X \sim N(\mu, \sigma^2)$. Then

$$1 - \Phi\left(\frac{3.287 - \mu}{\sigma}\right) = 0.063 ,$$

$$\Phi\left(\frac{3.287 - \mu}{\sigma}\right) = 0.937 = \Phi(1.53) .$$

Hence

$$3.287 - \mu = 1.53\sigma . \qquad (2)$$

Similarly

$$\Phi\left(\frac{2.897 - \mu}{\sigma}\right) = 0.488 = \Phi(-0.03) ,$$

giving

$$2.897 - \mu = -0.03\sigma . \qquad (3)$$

Subtracting (3) from (2) yields

$$\sigma = 0.25 .$$

Hence

$$\mu = 2.9045 .$$

Thus, the mean and variance of the distribution in question are 2.9045 and 0.0625.

6.5. The Central Limit Theorem

This is an important theorem which we shall not attempt to prove here. It states that, if $X_1, X_2, \ldots X_n$ are identically distributed with finite mean μ and variance σ^2, then the distribution of ΣX_i tends to $N(n\mu, n\sigma^2)$ as $n \to \infty$ (or equivalently, the distribution of $\overline{X}$ tends to $N(\mu, \sigma^2/n)$ as $n \to \infty$).

Certain particular applications of this theorem are the normal approximations to the binomial, Poisson, and gamma distributions. These are:

(i) If $K \sim B(p,n)$ where n is large and p is not too close to 0 or 1 (say $npq \geq 5$ as a rough guide), then K is approximately distributed as $N(np,npq)$.

(ii) If $K \sim Pn(m)$ where m is large, then K is approximately distributed as $N(m,m)$. Figure 4.2 shows a Poisson distribution approximating reasonably to the shape of a normal distribution, which illustrates the fact that even for m as low as 12, say, the normal approximation suffices.

(iii) If $X \sim Gm(a,b)$, then for large b, X is approximately distributed as $N(b/a, b/a^2)$.

The normal approximation to the binomial can carry over to a normal approximation for the hypergeometric distribution when s/N is small, but s is large and r/N not too near 0 or 1 (say $r(N-r)/N^2 \geq 5$). Then D, the number of 'specials' is approximately distributed as

$$N\left(\frac{rs}{N}, \frac{rs(N-r)(N-s)}{N^2(N-1)}\right).$$

There is a correction for continuity to be applied when the normal approximation is used for the distribution of a random variable which takes only integer values. Suppose the random variable is K with approximate distribution $N(\mu,\sigma^2)$. When we wish to calculate the probability of a set of values of k, for each k in that set, we must include the full interval $(k - \frac{1}{2}, k + \frac{1}{2})$. For example we have

$$P(K \leq k) \simeq \Phi\left(\frac{k + \frac{1}{2} - \mu}{\sigma} \right)$$

and

$$P(K < k) \simeq \Phi\left(\frac{k - \frac{1}{2} - \mu}{\sigma} \right).$$

Hence

$$P(K = k) \simeq \Phi\left(\frac{k + \frac{1}{2} - \mu}{\sigma} \right) - \Phi\left(\frac{k - \frac{1}{2} - \mu}{\sigma} \right)$$

$$\simeq \phi\left(\frac{k - \mu}{\sigma} \right),$$

where ϕ is the standard normal density function.

6.6. Reasons for the importance of the normal distribution

The use of the normal distribution is very prominent in modern statistical work for the following reasons:

(i) The normal distribution is found in practical experience to be appropriate (at least as a good approximation to the true distribution) for many types of physical measurements and other types of data.

(ii) The normal distribution and various related and derived distributions have been thoroughly studied and tabulated, so that it is very convenient, when data can reasonably be assumed to be normal, to use existing theory and tables.

(iii) Because of the Central Limit and related theorems a great many statistics used in statistical inference are approximately normally distributed for large sample sizes, whatever the original distribution of the data.

(iv) Many statistical techniques based upon the normal distribution are <u>robust</u>, that is, they remain approximately correct for reasonable departures from normality. Thus, for many purposes, little error is incurred by assuming normality, provided the departure from normality is not too great.

Despite these good reasons for assuming normality, we should warn against making this assumption without due thought and care. We shall see in Chapter 10 how to test the appropriateness of a particular distribution to given data.

7. The chi-squared distribution

A random variable X is said to be distributed as chi-squared with ν degrees of freedom if its density is given by

$$f(x) = x^{\frac{1}{2}\nu-1} e^{-\frac{1}{2}x}/\{\Gamma(\tfrac{1}{2}\nu) 2^{\frac{1}{2}\nu}\}, \qquad \nu > 0;\ 0 \leq x < \infty . \quad (1)$$

We shall denote this by writing $X \sim \chi^2_\nu$. An important example of such a random variable is the sum of squares of n independent standard normal variates which is distributed as χ^2_n; equivalently, if $X_1, X_2, \ldots X_n$ are independent random variables, each distributed as $N(\mu,\sigma^2)$, then $\Sigma(X_i-\mu)^2/\sigma^2$ is distributed as χ^2_n. It can also be shown that $\Sigma(X_i-\overline{X})^2/\sigma^2 = (n-1)s^2/\sigma^2$, where $\overline{X}$ is $\Sigma X_i/n$, is distributed as χ^2_{n-1}, independently of $\overline{X}$.

Figures 4.5 (a) and (b) show the densities of three chi-squared distributions; in (a), the special cases corresponding to $\nu = 1$ and 2 are shown while in (b), the more typical density curve is displayed. We note that the typical χ^2 distribution is skew. We also remark that the distribution function of χ^2 is widely tabulated since it plays an important role in statistical inference.

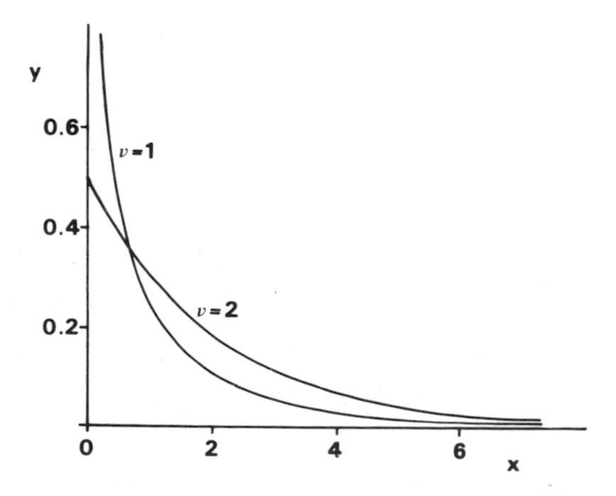

Figure 4.5. (a) $y = f(x)$, the density of $x = \chi^2$ for $\nu = 1,2$

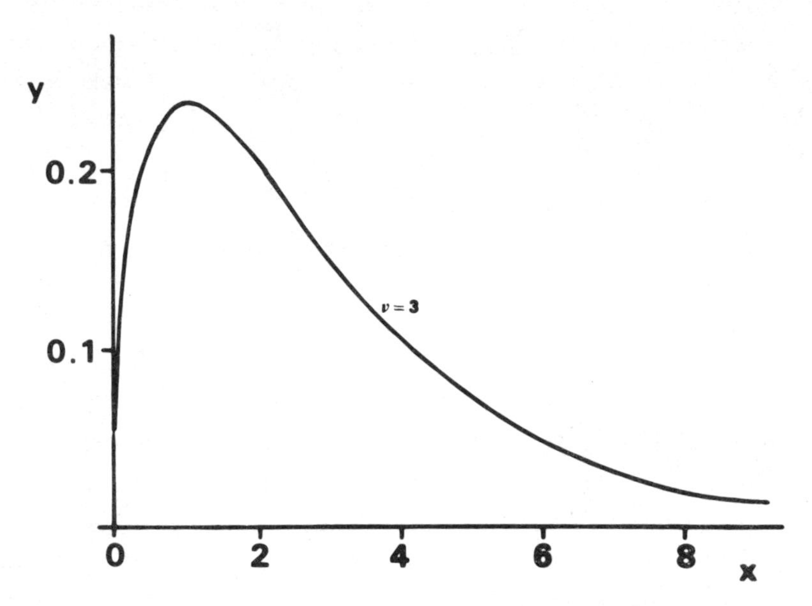

Figure 4.5. (b) $y = f(x)$, the density of $x = \chi^2$ for $\nu = 3$

7.1. Mean and variance

If $X \sim \chi^2_\nu$, the mean and variance are

$$E(X) = \nu ,$$

$$V(X) = 2\nu .$$

7.2. Additivity property

It can be shown using the methods of the previous sections that, if X_1 and X_2 are independently distributed as $\chi^2_{\nu_1}$ and $\chi^2_{\nu_2}$, then their sum, X_1+X_2 , is distributed as $\chi^2_{\nu_1+\nu_2}$.

7.3. Relationship to other distributions

The statement that X is distributed as χ^2_ν is equivalent to saying that $\frac{1}{2}X$ has the gamma distribution with $\nu/2$ degrees of freedom, that is, $\frac{1}{2}X \sim Gm(1,\nu/2)$ if $X \sim \chi^2_\nu$. Conversely, twice a gamma variate distributed as $Gm(1,\nu)$ has the chi-squared distribution with 2ν degrees of freedom. Moreover $\chi^2_\nu/\nu = F_{\nu,\infty}$ (see Section 9).

When ν is large, a χ^2_ν variate is approximately distributed as $N(\nu,2\nu)$. However, a better approximation for such a random variable is that $(2X)^{\frac{1}{2}}$ is approximately distributed as $N\{(2\nu-1)^{\frac{1}{2}},1\}$.

8. Student's t distribution

The random variable possessing this distribution is usually denoted by t, this being a case where by convention we do not distinguish between the random variable and its realization. The density of t is

$$f(t) = \frac{1}{\nu^{\frac{1}{2}}\beta(\frac{1}{2},\frac{1}{2}\nu)(1 + t^2/\nu)^{\frac{1}{2}(\nu+1)}}, \quad \nu > 0; -\infty < t < \infty . (1)$$

Here, ν is a constant, usually an integer, called the number of degrees of freedom of t. The other function, $\beta(\frac{1}{2},\frac{1}{2}\nu)$, appearing in equation (1) is the beta function which may be shown to be $\Gamma(\frac{1}{2}).\Gamma(\frac{1}{2}\nu)/\Gamma(\frac{1}{2}+\frac{1}{2}\nu)$. The beta and gamma functions are widely tabulated, but more important is the distribution function of t, a function which is to be found in the vast majority of statistical tables, because of its importance in tests on, and confidence intervals for, means of normal distributions (see Chapters 6 and 8).

A typical t distribution is shown in Figure 4.6. It can

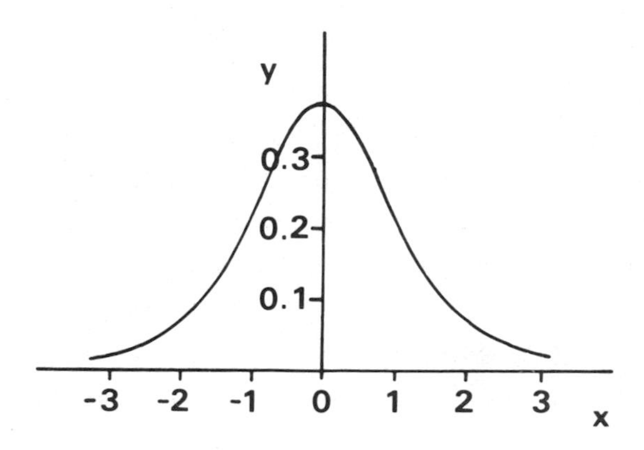

Figure 4.6. $y = f(x)$, the density of x = Student's t with $\nu = 4$

be seen that it is symmetrical and not unlike the normal distribution, though more squat.

8.1. Moments of the t distribution; mean and variance

From equation (1), it is clear that the density is symmetric about t = 0, so that all the odd moments which exist are zero, viz.

$$\mu_1' = \mu_3' = \mu_5' = \ldots = 0 .$$

Specifically, we observe the mean, μ_1', to be zero. The even moments are given by

$$\mu_{2r} = \frac{\nu^r \Gamma(r + \frac{1}{2}).\Gamma(\frac{1}{2}\nu - r)}{\Gamma(\frac{1}{2}).\Gamma(\frac{1}{2}\nu)} , \quad \text{for } 2r < \nu .$$

The moments, μ_r, only exist for $r < \nu$ so that the variance of t only exists if $\nu > 2$. From the above, we find

$$V(t) = \mu_2 = \nu/(\nu-2) .$$

8.2. Relationship to other distributions

If $Z \sim N(0,1)$ and $U \sim \chi^2_\nu$ independent of Z, then $T=Z/(U/\nu)^{\frac{1}{2}}$ is distributed as t with ν degrees of freedom.

There are two other quite general properties of interest. First, the square of a t_ν variate is an $F_{1,\nu}$ variate, where $F_{1,\nu}$ means the statistic F with $1,\nu$ degrees of freedom. This statistic is discussed in the next section. Secondly, as ν tends to infinity, the t distribution tends to normality; for example, the 97.5 percentile for $\nu \geq 28$, written as say $t_\nu(0.975)$, is 2.0 to one decimal place which is the same as $z(0.975)$ to the same number of significant figures.

9. The F distribution

A random variable is said to be distributed as F with degrees of freedom ν_1, ν_2, written as F_{ν_1,ν_2}, if its density is

$$\frac{\nu_1^{\frac{1}{2}\nu_1}\,\nu_2^{\frac{1}{2}\nu_2}\,x^{\frac{1}{2}(\nu_1-2)}}{\beta(\frac{1}{2}\nu_1,\frac{1}{2}\nu_2)(\nu_1 x+\nu_2)^{\frac{1}{2}(\nu_1+\nu_2)}}\,, \quad \nu_1,\ \nu_2 > 0; \quad 0 \leq x < \infty\,,$$

where the beta function $\beta(\frac{1}{2}\nu_1,\frac{1}{2}\nu_2) = \Gamma(\frac{1}{2}\nu_1)\Gamma(\frac{1}{2}\nu_2)/\Gamma(\frac{1}{2}\nu_1+\frac{1}{2}\nu_2)$. The density of a typical F distribution is illustrated in Figure 4.7. The most important use of the F distribution is

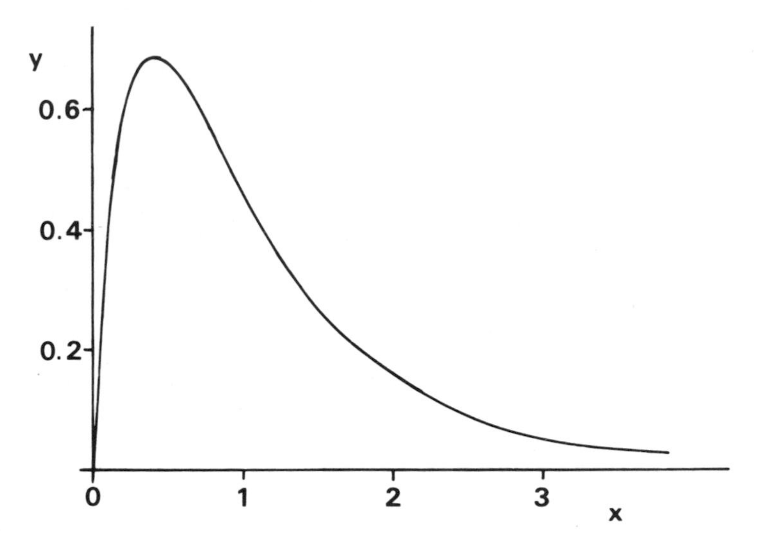

Figure 4.7. $y = f(x)$, the density of $x = F_{4,12}$

in tests involving the comparison of two distribution variances or in the procedure known as Analysis of Variance (see Chapters 8 and 9). For this purpose, values of certain percentiles of F are required and these are widely tabulated.

9.1. Mean and variance

It may be shown that

$$E(X) = \frac{\nu_2}{\nu_2 - 2} \quad \text{for } \nu_2 > 2\,,$$

and

$$V(X) = \frac{2\nu_2^2(\nu_1 + \nu_2 - 2)}{\nu_1(\nu_2 - 2)^2(\nu_2 - 4)} \quad \text{for } \nu_2 > 4\,.$$

9.2. Relationship to other distributions

If U and V are two random variables distributed independently as $\chi^2_{\nu_1}$ and $\chi^2_{\nu_2}$, then the ratio

$$\frac{U/\nu_1}{V/\nu_2}$$

is distributed as F_{ν_1,ν_2}. Clearly we could reverse the order of our two random variables and consider the ratio

$$\frac{V/\nu_2}{U/\nu_1} .$$

Obviously this ratio must be distributed as F_{ν_2,ν_1}, which, therefore, has the same distribution as $1/F_{\nu_1,\nu_2}$.

Note that, if $\nu_1 = 1$, we have $F_{1,\nu_2} = t^2_{\nu_2}$, and if $\nu_2 = \infty$, we have $F_{\nu_1,\infty} = \chi^2_{\nu_1}/\nu_1$.

10. Distribution of $\bar{X}$ and s^2 etc. in the normal case

It can be shown that, if $X_1, X_2, \ldots X_n$ are independently, identically distributed as $N(\mu,\sigma^2)$, then $\bar{X} \sim N(\mu,\sigma^2/n)$ and $\Sigma(X_i-\bar{X})^2/\sigma^2 \sim \chi^2_{n-1}$ independently of $\bar{X}$. Hence $(\bar{X}-\mu)/(\sigma/n^{\frac{1}{2}}) \sim N(0,1)$ and $(n-1)s^2/\sigma^2$ is independently distributed as χ^2_{n-1}, so that the ratio

$$\frac{\bar{X} - \mu}{s/n^{\frac{1}{2}}} \sim t_{n-1} .$$

We may extend this discussion to the case where we have two sets of data, $X_1, X_2, \ldots X_{n_1}$ independently identically distributed as $N(\mu_1,\sigma^2)$ and $Y_1, Y_2, \ldots Y_{n_2}$ independently identically distributed as $N(\mu_2,\sigma^2)$. Now $\bar{X} \sim N(\mu_1,\sigma^2/n_1)$ and $\bar{Y} \sim N(\mu_2,\sigma^2/n_2)$. Invoking the additivity property of the normal distribution shows that

$$\bar{X}-\bar{Y} \sim N\{\mu_1-\mu_2,\ \sigma^2(1/n_1 + 1/n_2)\}.$$

Hence

$$\frac{(\bar{X}-\bar{Y}) - (\mu_1-\mu_2)}{\sigma(1/n_1 + 1/n_2)^{\frac{1}{2}}} \sim N(0,1) .$$

We also know that $\Sigma(X_i-\bar{X})^2/\sigma^2 = (n_1-1)s_1^2/\sigma^2 \sim \chi^2_{n_1-1}$ and $\Sigma(Y_i-\bar{Y})^2/\sigma^2 = (n_2-1)s_2^2/\sigma^2 \sim \chi^2_{n_2-1}$. If we use the additivity property of χ^2, we find that

$$\frac{\Sigma(X_i-\bar{X})^2 + \Sigma(Y_i-\bar{Y})^2}{\sigma^2} \sim \chi^2_{n_1+n_2-2} .$$

Hence

$$\frac{(\bar{X}-\bar{Y}) - (\mu_1-\mu_2)}{(1/n_1 + 1/n_2)^{\frac{1}{2}}[\{\Sigma(X_i-\bar{X})^2 + \Sigma(Y_i-\bar{Y})^2\}/(n_1+n_2-2)]^{\frac{1}{2}}} \sim t_{n_1+n_2-2} .$$

This is useful when comparing two distribution means.

If we now suppose that the variance of the X's is σ_1^2 and that of the Y's is σ_2^2 , then the ratio

$$\frac{s_1^2/\sigma_1^2}{s_2^2/\sigma_2^2}$$

is distributed as F_{n_1-1,n_2-1} . In the special case where $\sigma_1^2 = \sigma_2^2$, the ratio of the two sample variances, s_1^2/s_2^2 , is distributed as F_{n_1-1,n_2-1} .

CHAPTER 5

ESTIMATION

In the previous two chapters, we have described some of the important parameters which characterize a distribution. It will be clear that none of these parameters can be obtained from a limited sample of data; all that can be done is to estimate them. Sometimes, we may directly estimate the density or distribution functions. The present chapter is concerned with the theory and practice of estimation.

1. The histogram

A histogram affords an estimation of a probability density function of a continuous random variable. We require a large number of observations, say, at least 50 but, better, at least 100. We count the number of data in each of several consecutive intervals, say, about 8 to 12 intervals. A figure is made with a column over each interval of area representing the count for that interval, so that the height of the column is the frequency (count) per unit x interval. If we were to gather more and more independent observations, and use more and narrower intervals, the histogram would tend nearer and nearer to a smooth curve of the same shape as the underlying density function. Indeed, if we were to scale the ordinates by dividing each count by the total number of observations, n, so that the area of each column represents the proportion of data lying in the corresponding interval, the histogram would tend towards coincidence with the graph of f(x), as n tends to infinity and the intervals become smaller. The use of area rather than height to represent the count or proportion ensures that this limiting coincidence is achieved. The use of area also has the advantage of giving a truer picture of the density function when the intervals are unequal.

Thus, a histogram gives us a good idea of what the underlying probability distribution of our observations is when we learn to recognize some of the more usual density shapes, and it affords a useful preliminary examination of a set of data of unknown distribution.

When the data have been measured and rounded off to a certain number of decimal places before allocation to the different intervals, the boundaries between consecutive intervals are defined with a 5 in the next decimal place. However, we could allocate data to different intervals as they are being observed, in which case the boundaries between intervals can be anything convenient and appropriate. For example, we could measure the lengths of a series of articles to two decimal places and allocate them to intervals (3.195, 3.395), (3.395, 3.595) etc; alternatively, we could use go/no-go gauges and, without actually measuring the lengths, count the articles in the intervals (3.2, 3.4), (3.4, 3.6), etc.

1.1. Example. The ages of 170 students were as shown in Table 5.1. The corresponding histogram is shown in Figure 5.1.

Table 5.1. Ages of a sample of students

Age/year	17	18	19	20	21	22	23	24
Count	2	30	45	51	27	10	4	1

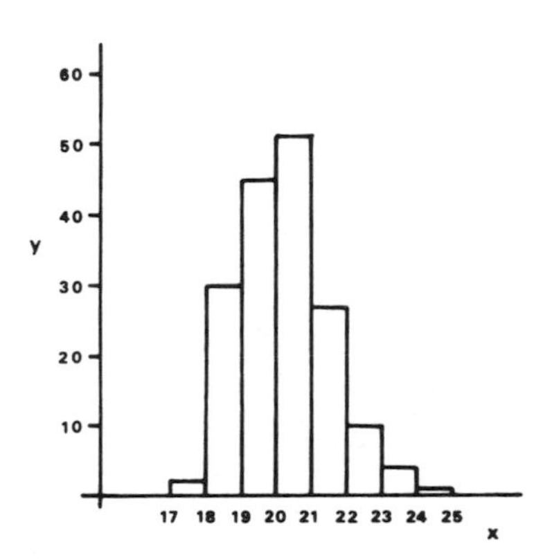

Figure 5.1. Histogram of students' ages, with y = frequency per unit x interval and x = age last birthday

2. The cumulative frequency function

The graph of this function (which is sometimes called an ogive) affords another diagram that is useful in the preliminary inspection of observations of a continuous random variable. Using counts of data per interval, we accumulate these counts consecutively to give the total number of data occurring below the upper boundary of each x interval. We then plot these cumulative totals against the upper limits and join adjacent points by straight lines. Interpolation using a linear portion between plotted points serves to estimate the cumulative frequency corresponding to a given x for the sample, and also (after scaling) to estimate F(x). If each interval tends to zero and the total number of data tends to infinity, the graph becomes of the same shape as that of the distribution function, F(x).

2.1. Example. We use the data of Section 1.1 to construct Table 5.2.

Table 5.2. Cumulative frequency of student ages

Age (upper limit)/year	18	19	20	21	22	23	24	25
Cumulative frequency	2	32	77	128	155	165	169	170

Thus, there are 32 students aged less than 19, 155 aged less than 22, etc.

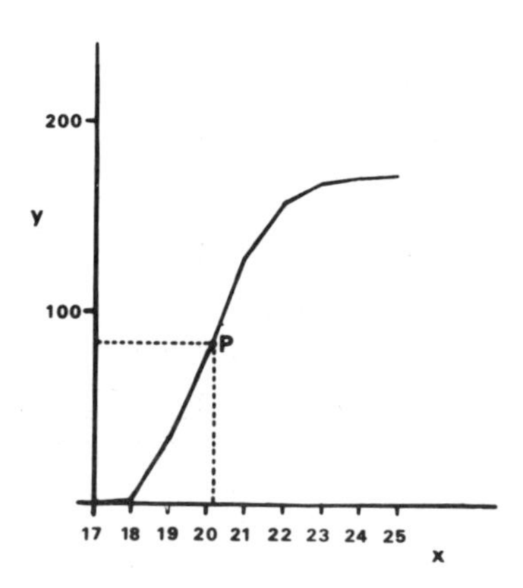

Figure 5.2. Cumulative frequency of students' ages with y = cumulative frequency and x = age. The point P is (20.2,85.5).

2.2. Estimation of a quantile, ξ_β, using cumulative frequency data

We may estimate ξ_β by reading off from the graph of the cumulative frequency function the x-value corresponding to $y = n\beta$, where n is the total number of data. A slightly better estimate of ξ_β is the value of x corresponding to $y = (n+1)\beta$, though clearly the difference between the two estimates is relatively small when n is large. For example, using Figure 5.2. the 0.5 quantile, or 50-percentile, (which is usually called the distribution median) is estimated, as shown, by the value 20.2. This latter value is an approximation to the sample median. When n is odd, the sample median is defined to be the middle value in order of size; when n is even, it lies between the two middle values and usually we take the average of these two as its value.

Alternatively, a quantile may be estimated without using the cumulative frequency curve by firstly noting in which interval the cumulative frequency of $(n+1)\beta$ lies and then taking the previous upper interval limit plus the appropriate proportion of the interval width as the estimate. For instance, using the data given in Table 5.2, the number $85.5 = (170 + 1)50/100$ corresponding to the median age of the students clearly lies in the interval (20,21) in which there are 51 data. We need (85.5 - 77) more data beyond the previous cumulative total of 77 and so, assuming that the 51 data are uniformly spread over the interval, we estimate the median age of the students as being the proportion (85.5 - 77)/51 of the way through the interval. Thus our new estimate is $20 + 8.5 \times 1/51 = 20.17$.

3. Estimators

In the previous two sections, we have been primarily concerned with obtaining some idea of the density and distribution functions to which our data conform in the favourable situation when a large number of data are available. Often we shall not be in a position to do this for the simple reason that it is not practicable to obtain the necessary large number of data. Nevertheless, we may still be able to estimate the parameters of the probability distribution if we have available a limited amount of data.

Suppose we have a set of independent random observations $X_1, X_2, \ldots X_n$. Suppose further that we can assume that these data conform to a particular probability distribution, but that there is a single unknown parameter, which we shall represent by θ, which is needed to fully define the density function. Our assumption may be based on theoretical considerations or on the form of the histogram of a large number of similar data. Each X has the density $f(x;\theta)$, say, or $p(x;\theta)$ if X is discrete, and distribution function $F(x;\theta)$. We estimate the parameter θ by a function of the X's, say T. Now T, which we call an <u>estimator</u>, is itself a random variable with its own probability distribution, mean, and variance. A particular value taken by T based on a particular set of data, that is, a <u>realization</u> of T, is called an <u>estimate</u>.

Some of the desirable properties of T are listed below.

(i) <u>Unbiasedness</u>

It is intuitively desirable that the distribution of T should be centred at θ, that is

$$E(T) = \theta . \tag{1}$$

An estimator possessing this property is called unbiased.

(ii) <u>Minimum variance unbiasedness</u>

The main aim is that T should have a high probability of being near to θ. There will usually be a number of alternative unbiased estimators, but some will have more widely spread distributions than others. Clearly such a large spread is undesirable because it means that the probability of T being close to its mean, θ, is small. Consequently, a desirable property, when it is attainable, is to have minimum variance among the class of unbiased estimators. Such an estimator is called <u>minimum-variance unbiased</u> (MVU), or <u>efficient</u>.

It can be shown that an MVU estimator exists if and only if the joint density of the independent observations, namely $f(x_1)f(x_2)\ldots f(x_n)$ if X is continuous or $p(x_1)p(x_2)\ldots p(x_n)$ if X is discrete is such that

$$\frac{\partial \ln L}{\partial \theta} = \lambda(\theta)(U-\theta) \ , \qquad (2)$$

where L is the joint density, called the likelihood, $\lambda(\theta)$ is some function of θ, possibly a constant, and U is some function of the X's alone. In this case, the MVU estimator will be U and its variance $1/\lambda(\theta)$. The MVU estimator when it exists, is unique. (Here, we are assuming that the range of x values for non-zero density is independent of θ).

(iii) Asymptotic minimum-variance unbiasedness

In many cases, we cannot obtain an MVU estimator. However, we can usually obtain an estimator which is asymptotically so, that is, the estimator is MVU for large sample size, n. Such estimators are useful when we have large n, and they will often be reasonably good even for small n.

(iv) Minimum mean square error

It is not really necessary for T to be unbiased provided that most of its probability distribution is near to θ; naturally, this requires that E(T) is near to θ. A measure of the tendency of T to be displaced from θ is $E\{(T-\theta)^2\}$, the mean square error, MSE. Thus, as an alternative to the first criterion mentioned, we might seek to find an estimator which has the minimum MSE among the set of possible estimators. Now, the MSE may be written

$$\begin{aligned} MSE &= E\{(T-\theta)^2\} \\ &= E\Big[[\{T-E(T)\} + \{E(T)-\theta\}]^2\Big] \\ &= V(T) + \{E(T)-\theta\}^2 \ , \qquad (3) \end{aligned}$$

as the cross-product term vanishes. Equation (3) shows that an MVU estimator seems to be a reasonable candidate for having minimum MSE. However, it can sometimes happen that an estimator with small bias, that is, small $E(T)-\theta$, has so small a variance as to have a lower MSE than any available MVU estimator.

(v) Sufficiency

If there is a function of the observations, say U, which contains all the information in the sample about the unknown parameter θ, then U is called a sufficient statistic for θ. Evidently, it is appropriate to use such a U, or some function of it, as an estimator of θ, or for any other inference concerning θ. For the record, we state that this property of sufficiency of U will be satisfied if and only if the conditional joint density of the observations given the value of U is independent of θ. It can be shown that this condition is fulfilled if and only if L factorizes into a function of the X's alone multiplied by a function of U and θ; equivalently, we must have

$$\frac{\partial \ln L}{\partial \theta} = \text{a function of } U \text{ and } \theta \text{ only.} \qquad (4)$$

In many cases, sufficient statistics do not exist, but they do for the parameters of the normal, Poisson, binomial, and exponential distributions. Comparing equations (2) and (4), we see that an MVU estimator is sufficient.

(vi) Consistency

It is clearly desirable that, for large n, T should be very close to θ. We should expect this since our estimator is based on so much information. This is the consistency property which, more specifically, is defined by saying that T is a consistent estimator of θ if

$$P(|T-\theta| < \varepsilon) \to 1 \text{ as } n \to \infty \text{ for any } \varepsilon > 0.$$

If T has zero bias (or a bias which tends to zero as n tends to infinity) and a variance which tends to zero as n tends to infinity, then T is consistent.

So far, we have been discussing the case of a single unknown parameter. In the more general case, parameters $\theta_1, \theta_2, \ldots$ would be estimated by random variables $T_1, T_2, \ldots$ Most of the above discussion concerning the desirable properties of estimators may be carried over to this case.

4. Some methods of estimation

A number of methods of estimation have been devised which satisfy some or all of the above criteria. A number of these are described below. We shall adopt the usual convention of representing _any_ estimator of a distribution parameter θ_i by the corresponding symbol 'starred', thus θ_i^*; the _maximum likelihood_ estimators, on the other hand, will be represented by the corresponding symbols 'hatted', thus $\hat{\theta}_i$.

(i) The method of maximum likelihood

This is probably the most used method of estimation, whereby θ is estimated by that value $\hat{\theta}$ which maximises L or, equivalently, ln L. Loosely speaking, we may say that we choose $\hat{\theta}$ so as to make the observations most likely, a criterion which clearly has an intuitive appeal. We take $\hat{\theta}$ as the solution of the likelihood equation,

$$\frac{\partial L}{\partial \theta} = 0 ,$$

or, equivalently and more usually,

$$\frac{\partial \ln L}{\partial \theta} = 0 , \qquad (1)$$

replacing realizations of the observations by the corresponding random variables. Clearly by comparing equations (3.2) with (1), we see that an MVU estimator, when it exists, will be given by this method; furthermore, it follows from equation (3.4) that, when a sufficient statistic exists, $\hat{\theta}$ will be a function of it and so will sufficient. Also $\hat{\theta}$ is consistent.

In most situations the reader is likely to meet, and provided that the range of non-zero values of the density is independent of θ, it can be shown that $\hat{\theta}$ is asymptotically MVU and, asymptotically distributed as

$$N\left\{\theta, -\frac{1}{E(\partial^2 \ln L/\partial\theta^2)}\right\} .$$

Moreover, the method is invariant, that is, the estimator, $\widehat{g(\theta)}$, of $g(\theta)$ for any function g with non-zero derivative is the same function of $\hat{\theta}$, thus

$$\widehat{g(\theta)} = g(\hat{\theta}) . \tag{2}$$

Note that, in order to find the expectation of $\hat{\theta}$, it is often helpful to make use of the relation

$$E\left(\frac{\partial \ln L}{\partial\theta}\right) = 0 .$$

The method may be used when we have r real parameters, $\theta_1, \theta_2, \ldots \theta_r$. In this case, we obtain r simultaneous equations,

$$\frac{\partial \ln L}{\partial\theta_j} = 0, \qquad j = 1 \text{ to } r,$$

which may ordinarily be solved to give the r estimators, $\hat{\theta}_1, \hat{\theta}_2, \ldots \hat{\theta}_r$. The properties (invariance, consistency, etc.) stated with regard to the single parameter apply to this case also.

The vector $\underline{\hat{\theta}}$ with ith element $\hat{\theta}_i$ will be asymptotically distributed as $N(\underline{\theta}, \underline{\Lambda}^{-1})$, where the ith element of $\underline{\theta}$ is θ_i and the (i,j)th element of $\underline{\Lambda}$ is

$$-E\left(\frac{\partial^2 \ln L}{\partial\theta_i \partial\theta_j}\right) .$$

(ii) The method of least squares

If we have $E(X_i) = g_i(\theta_1,\theta_2,...\theta_r)$, $i = 1$ to n, where g_i is a known function for each i, we may estimate $\theta_1,\theta_2,...\theta_r$ from a set of independent observations $X_1,X_2,...X_n$ by minimising the sum of squares, \$, given by

$$\$ = \Sigma(X_i - g_i)^2 ,$$

with respect to the θ's. This procedure gives r simultaneous equations

$$\frac{\partial \$}{\partial \theta_j} = 0, \qquad j = 1 \text{ to } r, \qquad (3)$$

whose solution gives the r estimators,

$$\theta_1^*,\theta_2^*,...\theta_r^* .$$

When the X's are normally distributed with the same variance, this method accords with the method of maximum likelihood. The method of least squares is most used for the case where each g_i is a linear combination of the θ's, a case which is dealt with more fully in Chapters 12 to 16. In this case, the method has relative computational simplicity.

(iii) The method of moments

If there are r real parameters, $\theta_1,\theta_2,...\theta_r$, we may estimate them by equating the first r distribution moments of X (which will be functions of $\theta_1,\theta_2,...\theta_r$) to the corresponding sample moments; that is, we write

$$E(X^j) = \sum_i X_i^j/n, \qquad j = 1 \text{ to } r. \qquad (4)$$

We solve these r equations for the r unknowns, giving $\theta_1^*,\theta_2^*,...\theta_r^*$. This method is invariant.

This procedure gives the same estimators of the mean and variance of a normal distribution as the method of maximum likelihood, but more generally it does not give the same estimators nor does it share the asymptotic efficiency.

(iv) The Bayesian method

This method is appropriate when it is possible to associate a prior probability distribution with the unknown parameter ; if, as is often the case, the Bayesian method is coupled with a decision theory approach, we also need a loss function which specifies the loss incurred as a result of the disparity between the parameter and its estimator. Our estimator is then that function of the observations which minimises the expectation of the loss (which is called the risk). However, we shall not pursue this method here.

4.1. Examples of the use of the method of maximum likelihood

Throughout this section, we shall assume that we have available n independent observations identically distributed, $X_1,X_2,\ldots X_n$. We shall examine three distinct situations. In the first two cases, we shall be concerned with the estimation of a single parameter only, whereas two quantities have to be estimated in the third case. We shall also consider in this section some aspects of the computation of the estimates.

Example 1. The X's are independently distributed as B(p,m) where m is known but p has to be estimated. Since

$$p(x) = {}^mC_x\, p^x(1-p)^{m-x} ,$$

it follows that

$$L = \left(\prod_i {}^mC_{x_i}\; p^{\Sigma x_i}\, (1-p)^{nm-\Sigma x_i}\right) .$$

Hence

$$\ln L = \Sigma \ln {}^mC_{x_i} + \Sigma x_i \ln p + (nm - \Sigma x_i)\ln(1-p) .$$

This gives

$$\frac{\partial \ln L}{\partial p} = \frac{\Sigma x_i}{p} - \frac{nm - \Sigma x_i}{1-p}$$

$$= \frac{nm}{p(1-p)}\left(\frac{\bar{x}}{m} - p\right) , \qquad (1)$$

where $\overline{x} = \Sigma x_i/n$. To obtain a maximum value for ln L, it follows from equation (1) that

$$\hat{p} = \frac{\overline{X}}{m} = \frac{\Sigma X_i}{nm} .$$

The expression for $\partial \ln L/\partial p$ has been deliberately cast into the form of equation (1) so as to correspond to equation (3.2). A comparison of these two equations shows that the maximum likelihood estimator, $\overline{X}/m$, is MVU with variance $p(1-p)/(nm)$. Furthermore, like all MVU estimators, $\hat{p}$ is sufficient.

As an example of these results, suppose we consider the compositions of families of four children with a view to estimating the probability that a child selected at random is a girl. Let X represent the number of girls so that

$$p(x) = {}^4C_x \, p^x(1-p)^{4-x} .$$

A survey of 12 of the authors' friends with four children gives the following realizations of X: 3, 2, 2, 0, 1, 3, 1, 1, 1, 2, 3, 1. Hence

$$\hat{p} = \frac{\Sigma x_i}{12 \times 4} = \frac{20}{48} = 0.42 .$$

This is quite close to the value of 0.5, which we might expect to be the true value of p.

Example 2. The X's are independently distributed as $Pn(\lambda)$ where λ has to be estimated. Since

$$p(x) = \frac{\lambda^x \exp(-\lambda)}{x!} ,$$

the likelihood is given by

$$L = \frac{\lambda^{\Sigma x_i} \exp(-n\lambda)}{\prod_i x_i!} .$$

Hence

$$\ln L = \Sigma x_i \ln \lambda - n\lambda - \Sigma \ln x_i!$$

Therefore

$$\frac{\partial \ln L}{\partial \lambda} = \Sigma x_i/\lambda - n$$

$$= \frac{n}{\lambda} (\bar{x} - \lambda) \ . \qquad (2)$$

Equation (2) shows that $\hat{\lambda} = \bar{X}$ which is MVU with variance λ/n; of course $\hat{\lambda}$ is also sufficient.

The data of Table 5.3 illustrate in a simple way the application of equation (2).

Table 5.3. Numbers of cosmic ray particles recorded in successive 1 second intervals.

106	106	111	100	93
115	130	98	103	95
125	94	120	99	93
117	103	108	125	100
123	92	101	125	115
109	105	108	84	147
102	104	133	106	111
142	122	115	92	103
120	106	126	138	108
96	124	104	113	120

These data are the number of cosmic ray particles picked up by a laboratory counter in successive 1 second intervals. On general grounds, we expect the number of such particles arriving in a fixed interval of time to be a random variable, say X, distributed as $Pn(\lambda)$ where λ is proportional to the length of the time interval, thus $\lambda = kt$. From the data in the table, we calculate

$$\bar{x} = 111.$$

Hence, an estimate of λ is

$$\hat{\lambda} = 111.$$

We saw in Chapter 3 that $s^2 = \Sigma(X_i-\bar{X})^2/(n-1)$ affords an unbiased estimator of the variance. Now, for a Poisson, the distribution variance is equal to the mean, λ. However, $\bar{X}$ is a better estimator of λ than s^2, being MVU. Here

$$s^2 = \lambda^* = 193$$

and, as we see, the two estimates do not agree all that well.

Example 3. The X's are independently distributed as $N(\mu,\sigma^2)$, where both μ and σ^2 are unknown and are to be estimated. Since

$$f(x) = (2\pi\sigma^2)^{-\frac{1}{2}} \exp\{-(x-\mu)^2/(2\sigma^2)\},$$

we obtain

$$L = (2\pi\sigma^2)^{-n/2} \exp\{-\Sigma(x_i-\mu)^2/(2\sigma^2)\} .$$

Hence

$$\ln L = -\frac{n}{2}\ln(2\pi\sigma^2) - \Sigma(x_i-\mu)^2/(2\sigma^2) . \qquad (3)$$

This gives two equations from which estimators for μ and σ^2 can be derived. These are

$$\frac{\partial \ln L}{\partial \mu} = 2\Sigma(x_i-\mu)/(2\sigma^2)$$

$$= \frac{n}{\sigma^2}(\bar{x}-\mu) = 0 \qquad (4)$$

and

$$\frac{\partial \ln L}{\partial \sigma^2} = -n/(2\sigma^2) + \Sigma(x_i-\mu)^2/(2\sigma^4) = 0 . \qquad (5)$$

Therefore, for maximum likelihood,

$$\hat{\mu} = \bar{X} . \tag{6}$$

Equation (4) also shows that $\hat{\mu}$ is MVU with variance σ^2/n. Equation (6) combined with equation (5) shows that

$$\hat{\sigma}^2 = \Sigma(X_i-\bar{X})^2/n . \tag{7}$$

The estimators $\hat{\mu}$ and $\hat{\sigma}^2$ are jointly sufficient for μ and σ^2.

It is very easy to show that

$$\Sigma(X_i-\bar{X})^2 = \Sigma X_i^2 - n\bar{X}^2 . \tag{8}$$

Hence an alternative form of equation (7) is

$$\hat{\sigma}^2 = \overline{X^2} - \bar{X}^2 ;$$

that is, $\hat{\sigma}^2$ is the difference between the mean square and the squared mean. A second form of equation (8) is particularly useful for computational purposes:

$$\Sigma(X_i-\bar{X})^2 = \Sigma X_i^2 - (\Sigma X_i)^2/n . \tag{9}$$

The use of equation (9) avoids introducing round-off error as a result of calculating the differences $(x_i-\bar{x})$ with $\bar{x}$ expressed to too few significant figures. Furthermore, experience shows that the use of the right-hand side of equation (9) leads to far fewer gross errors than the use of the left-hand side. It is also much more convenient when a calculating machine with at least two storage registers is being used since then only one entry of each datum needs to be made.

It can be shown that

$$E(\hat{\sigma}^2) = \frac{(n-1)\sigma^2}{n} ,$$

and so $\hat{\sigma}^2$ is biased. For this reason, the unbiased estimator

$$s^2 = \Sigma(X_i-\bar{X})^2/(n-1) \qquad (10)$$

is usually used for σ^2. Clearly s^2 and $\hat{\sigma}^2$ are close to each other for large n.

As a simple example of the calculation of estimates of mean and variance, we use the data of Table 5.4. These data refer to five samples of

Table 5.4. Densities of five samples of a high molecular weight polymer, H1.

Sample	1	2	3	4	5
Density/g cm^{-3}	1.2151	1.2153	1.2155	1.2145	1.2151

a high molecular weight polymer, H1. The five samples were taken from the same batch and their densities, ρ, were measured using a flotation method. As usual, we define our random variable, X, to be dimensionless, in this case through the expression

$$X = \frac{\rho}{g\ cm^{-3}} .$$

Proceeding straightforwardly, we obtain

$$\Sigma x_i^2 = 7.38234061$$

$$(\Sigma x_i)^2/n = 7.38234005$$

$$\bar{x} = 1.21510$$

$$\hat{\sigma}^2 = 11.2 \times 10^{-8} \qquad \hat{\sigma} = 3.3 \times 10^{-4}$$

$$s^2 = 14.0 \times 10^{-8} \qquad s = 3.7 \times 10^{-4}$$

Note the large number of significant figures which have to be carried in this calculation, clearly a potential source of trouble.

A much better procedure is to code the data by a suitable linear transformation of the type

$$y = ax + b ,$$

where y represents the transformed quantity, and a and b are chosen so as to reduce the number of significant figures required for y without at the same time losing any of the information contained in the data. Now it follows immediately that

$$\bar{y} = a\bar{x} + b \qquad \bar{x} = (\bar{y}-b)/a$$

and

$$s^2(y) = a^2 s^2(x) \qquad s^2(x) = s^2(y)/a^2 \ .$$

Obvious choices for a and b are 10^4 and -12151 in this case. We thus obtain the data of Table 5.5.

Table 5.5. The coded data of Table 5.4

x	y	y^2
1.2151	0	0
1.2153	2	4
1.2155	4	16
1.2145	-6	36
1.2151	0	0
sums	0	56

From the data of Table 5.5 we deduce

$$\Sigma y_i^2 = 56 \ ,$$

$$(\Sigma y_i)^2/n = 0 \ ,$$

$$\bar{y} = 0 \longrightarrow \bar{x} = 1.21510 \ ,$$

$$\hat{\sigma}^2(y) = 11.2 \longrightarrow \hat{\sigma}^2(x) = 11.2 \times 10^{-8} \ ,$$

$$s^2(y) = 14.0 \longrightarrow s^2(x) = 14.0 \times 10^{-8} \ .$$

Clearly the probability of error is greatly reduced by this procedure. Returning now to polymer density, we conclude that the maximum likelihood estimates of the mean, variance, and standard deviation are 1.21510g cm^{-3}, 11.2×10^{-8} g^2 cm^{-6}, and 3.3×10^{-4} g cm^{-3}. The standard error of $\bar{\rho}$, $s(\bar{\rho}) = s(\rho)/n^{\frac{1}{2}}$, is 1.7×10^{-4} g cm^{-3}.

With a large quantity of grouped data in intervals of equal width, we usually estimate μ and σ^2 by calculating suitable approximations to $\bar{X}$ and $\Sigma(X_i-\bar{X})^2/n$. The approximation is effected by treating the data in each interval as if they were at the centre of the interval; we thereby greatly reduce the amount of calculation without incurring any serious error in the estimation of μ and σ^2.

We illustrate the procedure with the following data, derived from observations of the time of transit of Polaris over the Greenwich meridian using a clock measuring sidereal time. In this work, the deviation, t, of the observed transit time from a value somewhere near the mean of the data was recorded, together with f, the frequency with which t fell into the following intervals (-3.75, -3.25) seconds, (-3.25, -2.75) seconds, etc. For convenience, we tabulate the values of f against the mid-points of the intervals, as shown in Table 5.6.

Table 5.6. Deviations of the transit times of Polaris at the Greenwich meridian from a value near the mean

t/s	f	t/s	f
-3.5	2	0	168
-3.0	12	0.5	148
-2.5	25	1.0	129
-2.0	43	1.5	78
-1.5	74	2.0	33
-1.0	126	2.5	10
-0.5	150	3.0	2

As in the simpler example, we code the data by a suitable linear transformation. In this case, the data are already centralised around zero so that very little is needed to simplify the arithmetic; an appropriate transformation, here, is

$$y = \frac{2t}{\text{second}} ,$$

where t stands for the mid-point of an interval. This gives us the column of (dimensionless) integers, shown as column 3 in Table 5.7, and greatly

Table 5.7. Computational table, including check columns, from the data of Table 5.6

Col: 1	2	3	4	5	6	7	8	9
t/s	f	y	fy	fy^2	y+1	f(y+1)	$f(y+1)^2$	cum.f
-3.5	2	-7	-14	98	-6	-12	72	2
-3.0	12	-6	-72	432	-5	-60	300	14
-2.5	25	-5	-125	625	-4	-100	400	39
-2.0	43	-4	-172	688	-3	-129	387	82
-1.5	74	-3	-222	666	-2	-148	296	156
-1.0	126	-2	-252	504	-1	-126	126	282
-0.5	150	-1	-150	150	0	-575	0	432
0	168	0	-1007	0	1	168	168	600
0.5	148	1	148	148	2	296	592	748
1.0	129	2	258	516	3	387	1161	877
1.5	78	3	234	702	4	312	1248	955
2.0	33	4	132	528	5	165	825	988
2.5	10	5	50	250	6	60	360	998
3.0	2	6	12	72	7	14	98	1000
	1000		834	5379		1402	6033	
			-1007			-575		
			-173			827		

simplifies the arithmetic. Column 4 is obtained by multiplying corresponding terms in columns 2 and 3. Column 5 of this table is most easily obtained by multiplying corresponding terms in columns 3 and 4; similarly column 8 is obtained from columns 6 and 7. Here columns 6, 7, and 8 are optional extras, introduced to check the calculations as follows:

$$\text{(i)} \qquad \Sigma f(y+1) = \Sigma fy + \Sigma f = -173 + 1000 = 827 \;, \quad \checkmark$$

$$\text{(ii)} \qquad \Sigma f(y+1)^2 = \Sigma fy^2 + 2\Sigma fy + \Sigma f = 5379 - 346 + 1000 = 6033 \;. \quad \checkmark$$

In columns 4 and 7, the negative values have been subtotalled for convenience where a central zero value would occur.

From the totals shown in Table 5.7, we obtain

$$\bar{y} = \Sigma fy/n = -173/1000 = -0.173$$

and

$$\hat{\sigma}^2(y) = \Sigma f(y-\bar{y})^2/n = \overline{y^2} - (\bar{y})^2 = 5.379 - (0.173)^2 = 5.349 \;,$$

giving

$$\hat{\sigma}(y) = 2.3 \;.$$

To transform back to the variable t, we note that t = y second/2, so that

$$\bar{t} = \bar{y}\ \text{second}/2$$

and

$$\hat{\sigma}^2(t) = \hat{\sigma}^2(y)\text{second}^2/4 \;.$$

Hence our final values are $\bar{t} = -0.087$ second, $\hat{\sigma}^2(t) = 1.3$ second^2, $\hat{\sigma}(t) = 1.2$ second, and $\hat{\sigma}(\bar{t}) = 0.037$ second.

It is convenient for us at this point to illustrate the calculation of the sample median from grouped data. Column 9 of Table 5.7 gives the cumulative frequencies at the upper ends of the intervals. We quickly see that the value of the median t lies between -0.25 second and +0.25

second (remember that column 1 gives the mid-points of the intervals). Since the median corresponds to a cumulative frequency of $(1000 + 1)/2 = 500.5$, we estimate its position by simple proportion, thus

$$(-0.25 + \frac{500.5 - 432}{168} \times 0.5)\text{second} = -0.046 \text{ second.}$$

Of course, we have had to assume a uniform density for t over the interval (-0.25, 0.25) second, but this assumption will not usually be too far out.

A second example of this computational procedure is given in Chapter 10, Table 10.2.

4.1.1. Number of significant figures for a sample mean, sample standard deviation, and standard error of the mean

We digress at this point to deal with a common problem, namely how many significant figures shall we quote following the calculation of a sample mean and sample standard deviation. Now, the distribution variance of a sample standard deviation for a set of normally distributed data is $\sigma^2/2n$ so that the coefficient of variation, is $(\sigma^2/2n)^{\frac{1}{2}}/\sigma = (1/2n)^{\frac{1}{2}}$. The coefficient of variation, c.v., of a random variable is the ratio of its standard error to its mean; the c.v. is often expressed as a percentage. Hence, even with n as large as 50, the c.v. is 10 percent and, for it to be reduced to 1 per cent, n would have to be as large as 5,000. In view of this, it would hardly be appropriate to record more than two significant figures in the sample standard deviation. One could argue that the spread of the distribution of the sample standard deviation is so large that 2 significant figures for the quantity is too many and that 1 would serve better. However, since further calculations are often carried out with the sample standard deviation, it is important not to discard too many digits too early else rounding off errors will occur. It is for this reason that we advocate the presentation of two digits for s. To achieve two figure accuracy in s, it is safer to keep three significant figures for the sample variance. We would actually carry rather more significant figures during the calculation of these two quantities as we often lose some in the process; we also wish to avoid excessive accumulation of rounding-off errors. It is not possible to give hard and fast rules as to how many digits are retained during a calculation since computations differ so much one from another.

If two significant figures are appropriate for the sample standard deviation, two significant figures are appropriate for the standard error of the mean, $s/n^{\frac{1}{2}}$. This gives us a basis for deciding to how many significant figures we quote the sample mean. We evaluate $s(\bar{x})$ and then quote our final value of the sample mean to the same number of decimal places as the second significant figure of the value of $s(\bar{x})$. This gives us a tidy result and avoids excessive accumulation of rounding-off errors in any subsequent calculations.

Although our arguments have been based on a normally distributed set of data, they may be expected to apply reasonably well to other types of distribution.

We shall always separately name or symbolize our estimates and <u>not</u> write them in the form

$$1.21510 \pm 0.00017 ,$$

because this has several possible interpretations.

4.2. An example of the use of the method of least squares

As before, we have a set of observations, $X_1, X_2, \ldots X_n$, which are independently, identically distributed. We wish to estimate the mean, μ, using the method of least squares. Put

$$\$ = \Sigma(X_i - \mu)^2$$

so that

$$\frac{\partial \$}{\partial \mu} = -2\Sigma(X_i - \mu)$$

and

$$\frac{\partial^2 \$}{\partial \mu^2} = 2n .$$

Hence \$ is a minimum when $\mu = \bar{X}$ which is thus our least squares estimator. This is $\hat{\mu}$ when the data are normal. Indeed in this case, the method of least squares is simply a variant of the method of maximum likelihood since the minimization of \$ is exactly what is required to maximise ln L (see equation 4.1.3).

4.3. Some results from the method of moments

The method of moments estimates μ by $\bar{X}$ and σ^2 by $\Sigma(X_i-\bar{X})^2/n$ for any distribution of the X's which has a mean and variance. However, we shall not exemplify the method of moments in detail. We simply remark that the method of moments gives the same estimators of the parameters p, λ, μ and σ^2 of the binomial, Poisson, and normal distributions as those obtained by the method of maximum likelihood.

5. Estimation of the mean and variance of a function

Suppose we have a function, ϕ, of the r physical quantities $\theta_1, \theta_2, \ldots \theta_r$ which we estimate severally by unbiased estimators $T_1, T_2, \ldots T_r$. We require to estimate $\phi(\theta_1, \theta_2, \ldots \theta_r)$ and its standard deviation. We remark in passing that the estimated standard deviation of a quantity which is derived from the original observations is often termed its standard error. The obvious estimator to use is $\phi(T_1, T_2, \ldots T_r)$, which, as we shall show, is approximately unbiased with variance given by

$$V\{\phi(T_1, T_2, \ldots T_r)\} \approx \sum_i \left(\frac{\partial \phi}{\partial \theta_i}\right)^2 V(T_i) + 2 \sum_{i<j} \left(\frac{\partial \phi}{\partial \theta_i}\right)\left(\frac{\partial \phi}{\partial \theta_j}\right) C(T_i, T_j). \quad (1)$$

Here, the partial derivatives are evaluated at $\{\theta_i\}$ and the second summation is the sum over all i,j = 1,2,...r such that $i < j$. Note that, if ϕ consists of products of powers (some negative, possibly) of the T's, it is usually easiest to apply equation (1) to ln ϕ rather than to ϕ itself.

To prove these statements, we use Taylor's theorem to obtain

$$\phi(T_1,T_2,\ldots T_r) = \phi(\theta_1,\theta_2,\ldots\theta_r) + \sum\left(\frac{\partial\phi}{\partial\theta_i}\right)(T_i-\theta_i)$$

$$+ \text{ higher order terms.} \qquad (2)$$

If the variances of the T's are relatively small, these higher order terms are small compared with the leading terms of equation (2) and so may be ignored. Hence

$$\phi(T_1,T_2,\ldots T_r) \approx \phi(\theta_1,\theta_2,\ldots\theta_r) + \sum\left(\frac{\partial\phi}{\partial\theta_i}\right)(T_i-\theta_i) \ . \qquad (3)$$

Now $E(T_i-\theta_i) = 0$, since the T's are unbiased. Therefore, taking expectations of both sides of equation (3) gives

$$E\{\phi(T_1,T_2,\ldots T_r)\} \approx \phi(\theta_1,\theta_2,\ldots\theta_r) \ ,$$

and so our estimator is approximately unbiased.

For the variance of our estimator, we write

$$V\{\phi(T_1,T_2,\ldots T_r)\} = E\left[\{\phi(T_1,T_2,\ldots T_r) - \phi(\theta_1,\theta_2,\ldots\theta_r)\}^2\right] \ ,$$

which, from equation (3), becomes

$$V\{\phi(T_1,T_2,\ldots T_r)\} = E\left[\left\{\sum\left(\frac{\partial\phi}{\partial\theta_i}\right)(T_i-\theta_i)\right\}^2\right] \ .$$

On expansion of the right-hand side, we obtain equation (1). If, as is often the case, the T's are independent, then $C(T_i,T_j) = 0$, all i,j, and equation (1) simplifies to

$$V\{\phi(T_1,T_2,\ldots T_r)\} = \sum\left(\frac{\partial\phi}{\partial\theta_i}\right)^2 V(T_i) \ . \qquad (4)$$

In practice, each $V(T_i)$ would usually be replaced by its estimator $s^2(T_i)$, each $C(T_i,T_j)$ by its estimator, $\hat{C}(T_i,T_j)$, and the partial derivatives evaluated at $\theta_i = t_i$, the realization of T_i, for $i = 1,2,\ldots r$. Hence our estimator of $V\{\phi(T_1,T_2,\ldots T_r)\}$ is

$$s^2\{\phi(T_1,T_2,\ldots T_r)\} = \begin{cases} \sum_i \left(\frac{\partial\phi}{\partial\theta_i}\right)^2 s^2(T_i) + 2\sum_{i<j}\left(\frac{\partial\phi}{\partial\theta_i}\right)\left(\frac{\partial\phi}{\partial\theta_j}\right)\hat{C}(T_i,T_j) , \\ \text{or } \sum_i \left(\frac{\partial\phi}{\partial\theta_i}\right)^2 s^2(T_i) \text{ for independent T's.} \end{cases} \quad (5)$$

Just as we would use the unbiased estimator

$$s^2(T_i) = \sum_k (T_{ik}-\bar{T}_i)^2/(n-1)$$

for a set of n data (see Chapter 3, Section 4), so we would use the unbiased estimator

$$\hat{C}(T_i,T_j) = \sum_k (T_{ik}-\bar{T}_i)(T_{jk}-\bar{T}_j)/(n-1) .$$

We may wish to have a value for the number of degrees of freedom associated with $s^2\{\phi(T_1,T_2,\ldots T_r)\}$ when the $s^2(T)$'s are associated with different numbers of degrees of freedom. A suitable approximation (Satterthwaite, 1946) is

$$\bar{\nu} = \frac{s^4\{\phi(T_1,T_2,\ldots T_r)\}}{\sum\left(\frac{\partial\phi}{\partial\theta_i}\right)^4 s^4(T_i)/\nu_i} , \quad (6)$$

where ν_i is the number of degrees of freedom associated with $s^2(T_i)$.

Example 1. As a simple illustration of the use of equation (4), suppose we were to determine the density, D, of a ball-bearing by measuring (independently) its mass, M, and radius, R. Now

$$D = \frac{3M}{4\pi R^3} .$$

We take logarithms of both sides of the above equation, as advised earlier when discussing the general result, equation (1), to obtain

$$\ln D = \ln\{3/(4\pi)\} + \ln M - 3\ln R .$$

Hence, from equation (4),

$$\frac{V(D)}{D^2} = \frac{V(M)}{M^2} + \frac{9V(R)}{R^2} ,$$

where V(M) and V(R) are the variances of M and R. It follows immediately that

$$\frac{s^2(D)}{d^2} = \frac{s^2(M)}{m^2} + \frac{9s^2(R)}{r^2} ,$$

where $s^2(D)$, $s^2(M)$, and $s^2(R)$ are estimated variances of D, M, and R, m and r are the recorded values of mass and radius, and d is the estimated density.

Example 2. The focal length, f, of a thin convex lens is given by the formula

$$\frac{1}{f} = \frac{1}{u} + \frac{1}{v} ,$$

where u and v stand for the object and image distances. In an experiment, u and v were measured to the nearest mm and recorded as 35.1 and 41.1 cm respectively. Our problem is to estimate f and its standard error.

To estimate f, we simply substitute the realizations of u and v in the above equation.

$$f = \left(\frac{1}{35.1} + \frac{1}{41.1}\right)^{-1} \text{ cm}$$

$$= 18.9319 \text{ cm to four decimal places.}$$

To find the standard error of f, we have to have values for the variances of u and v. We assume that the recorded values of u and v are subject only to rounding-off error within ± 0.05 cm. In this case, we may use the result derived in Section 3.2 of Chapter 3 to obtain

$$V(u) = V(v) = 8.33 \times 10^{-4} \text{ cm}^2 .$$

Now u and v are independent so that we may use equation (4) to obtain V(f), thus

$$V(f) = \left(\frac{\partial f}{\partial u}\right)^2 V(u) + \left(\frac{\partial f}{\partial v}\right)^2 V(v) .$$

Further

$$-\frac{1}{f^2}\left(\frac{\partial f}{\partial u}\right) = -\frac{1}{u^2} ,$$

so that

$$\left(\frac{\partial f}{\partial u}\right) = \frac{f^2}{u^2} ,$$

and similarly for $(\partial f/\partial v)$. Hence

$$V(f) = f^4\{V(u)/u^4 + V(v)/v^4\}$$

$$= 0.0001080 \text{ cm}^2$$

and

$$\sigma(f) = 0.0104 \text{ cm}.$$

It only remains to decide the number of significant figures to be associated with $\sigma(f)$ and f. Two significant figures are quite adequate for $\sigma(f)$ in view of the approximations involved; as a result, we should only quote f to three decimal places

$$f = 18.932 \text{ cm} ,$$

$$\sigma(f) = 0.010 \text{ cm} .$$

We remark that, although the rounding-off errors in u and v are uniformly distributed, the resulting error in f is not (see next section).

6. Deterministic and statistical errors due to rounding-off errors

As in the last section, we have a quantity $\phi(\theta_1,\theta_2,\ldots\theta_r)$ estimated by the quantity $\phi(T_1,T_2,\ldots T_r)$, where the true values of the parameters $\theta_1,\theta_2,\ldots\theta_r$ are estimated by the observations $T_1,T_2,\ldots T_r$ and the differences $T_1-\theta_1$, $T_2-\theta_2$, etc. are random errors. These errors may be experimental, or they may be due to rounding-off, or they may be of both types. Here we shall consider errors due only to rounding-off.

We use equation (5.2) neglecting again the second-order and higher terms to consider the overall error in $\phi(T_1, T_2, \ldots T_r)$. This is

$$\phi(T_1, T_2, \ldots T_r) - \phi(\theta_1, \theta_2, \ldots \theta_r) = \sum\left(\frac{\partial \phi}{\partial \theta_i}\right)(T_i - \theta_i) \quad . \qquad (1)$$

We let e_i represent the maximum value of $|T_i - \theta_i|$; this will usually be 0.5 in units of the last decimal place used for recording t_i. Using equation (1), the maximum possible overall error (the deterministic error) will be given by

$$|\phi(T_1, T_2, \ldots T_r) - \phi(\theta_1, \theta_2, \ldots \theta_r)| \leq \sum \left|\left(\frac{\partial \phi}{\partial \theta_i}\right)\right| e_i = M, \text{ say.} \qquad (2)$$

Usually, however, the actual overall error will be much less than M, for the contributions to the right-hand side of equation (1) will be of both signs and will tend to compensate each other. Accordingly, we look for a statistical error sometimes called a stochastic error below which the overall error in $\phi(T_1, T_2, \ldots T_r)$ will probably fall. In general, this will be much smaller than M, but more realistic.

Equation (1) shows that the error in $\phi(T_1, T_2, \ldots T_r)$ is the sum of a number of random variables. If r is large and the T's are independent, we may invoke the Central Limit Theorem to show that this error is approximately normally distributed. Specifically, we may write

$$E = \phi(T_1, T_2, \ldots T_r) - \phi(\theta_1, \theta_2, \ldots \theta_r) \sim N\left\{0, \sum\left(\frac{\partial \phi}{\partial \theta_i}\right)^2 V(T_i)\right\}$$

from the results derived in Section 5, where E stands for the error. If for each i, we suppose that $(T_i - \theta_i)$ has a rectangular distribution between $\pm\, e_i$, we may substitute $e_i^2/3$ for $V(T_i)$ in the above expression for the variance of the stochastic error (see Chapter 3, Section 3.2). Hence

$$E \sim N\left\{0, \sum\left(\frac{\partial \phi}{\partial \theta_i}\right)^2 e_i^2/3\right\} .$$

Although r is not usually large, the fact that the errors are all symmetrically distributed will improve the approximate normality. We know that, in the case of a normally distributed variate, about 95% of the observations lie within ± two standard deviations about the mean. Hence, 95% of the errors will be within the range

$$0.577 \left\{ \sum \left(\frac{\partial \phi}{\partial \theta_i} \right)^2 (2e_i)^2 \right\}^{\frac{1}{2}} ,$$

where $0.577 = 1/3^{\frac{1}{2}}$. This is often called the stochastic error.

CHAPTER 6

CONFIDENCE INTERVALS

We have already stated in the last chapter that a desirable estimator is one which has a high probability of being near to the unknown parameter which it estimates. Clearly, we should like to quantify this by evaluating the probability that our estimator lies within a certain range of the parameter in question. Usually, we think in terms of a range centred on the parameter so that, if T is our estimator and θ is the quantity which it estimates, we have to evaluate the probability that T will lie in the interval $\theta \pm e$, where e is a constant.

The statement that T lies in $\theta \pm e$ is equivalent to saying that θ lies in $T \pm e$, which thus has the same probability. However, in this second statement, θ is a constant and the interval is random, whereas in the first statement it is T which is random and the interval which is constant (for a given e). A realization of such a random interval is called a confidence interval and the probability that the random interval includes the true value of θ is called the confidence coefficient.

More generally, we can calculate two limits from our sample of observations, $L_1(X_1,X_2,...X_n)$ and $L_2(X_1,X_2,...X_n)$ where $L_1 < L_2$, and

$$P(L_1 \leq \theta \leq L_2) = 1 - \alpha .$$

Here α is small, usually 0.05, 0.01, or 0.001, in descending order of popularity. Then, $1-\alpha$ is called a confidence coefficient and any realization, (ℓ_1,ℓ_2), of the random interval, (L_1,L_2), is called a confidence interval. Thus a confidence interval does not have to be symmetrically disposed about the true value though this is a common situation. We cannot say that a confidence interval (ℓ_1,ℓ_2) has a probability of $1-\alpha$ of including the true value of θ since with particular values of ℓ_1 and ℓ_2 and a constant θ, the probability that (ℓ_1,ℓ_2)

includes θ is either 1 or 0 according to whether it does or does not include θ. What we can say is that we have a <u>confidence</u> of $1-\alpha$ that (ℓ_1,ℓ_2) includes the true value of θ, because, if we draw a large number of samples of size n and calculate (ℓ_1,ℓ_2) for each, then approximately a proportion $1-\alpha$ would contain θ.

A confidence interval affords more information about the unknown θ than an estimate, because the confidence coefficient and the interval width give us an indication of how close to θ we are.

We shall now illustrate the derivation of confidence intervals associated with the means and variances relating to the normal distribution.

1. <u>Interval for mean</u>

We deal here with three cases, distinguished from each other according to the information we have about the distribution of the observations.

<u>Case 1</u>. Normal distribution with known variance

Although this case is exceptional, it provides a simple illustration of the method of computing a confidence interval and, at the same time, it paves the way for the more usual case, Case 2. It arises, for instance, when a large amount of preliminary data have been accumulated for which we have calculated the mean and variance; because of the large sample size, we may take these values as being virtually equal to the corresponding distribution values. Now if our particular sample of observations has been obtained under some changed conditions which are likely to affect only the mean, but not the variance, then this case applies.

Suppose our observations are $X_1,X_2,\ldots X_n$ independently identically distributed as $N(\mu,\sigma^2)$, where μ is unknown but σ^2 is known. In this case, we have seen that the sample mean $\bar{X} \sim N(\mu,\sigma^2/n)$, so that

$$P(\bar{X}-e \leq \mu \leq \bar{X}+e) = P(\mu-e \leq \bar{X} \leq \mu+e)$$

$$= P\left(\frac{-e}{\sigma/n^{\frac{1}{2}}} \leq \frac{\bar{X}-\mu}{\sigma/n^{\frac{1}{2}}} \leq \frac{e}{\sigma/n^{\frac{1}{2}}}\right) ,$$

where e is any constant and $(\bar{X}-\mu)/(\sigma/n^{\frac{1}{2}})$ is the standardised form of $\bar{X}$. Hence

$$P(\bar{X}-e \leq \mu \leq \bar{X}+e) = \Phi\left(\frac{e}{\sigma/n^{\frac{1}{2}}}\right) - \Phi\left(\frac{-e}{\sigma/n^{\frac{1}{2}}}\right)$$

$$= 2\Phi\left(\frac{e}{\sigma/n^{\frac{1}{2}}}\right) - 1 , \qquad (1)$$

where Φ is the standard normal distribution function. We must choose e so that the right-hand side of equation (1) is equal to our chosen value of $1-\alpha$. Hence

$$\Phi\left(\frac{e}{\sigma/n^{\frac{1}{2}}}\right) = 1 - \alpha/2 ,$$

which fixes e completely when σ and n are known. If we write $z(1-\alpha/2)$ in place of $e/(\sigma/n^{\frac{1}{2}})$ to signify the standard normal variate corresponding to $\Phi\{z(1-\alpha/2)\} = 1-\alpha/2$, the interval for μ with coefficient $1-\alpha$ is

$$\bar{x} \pm z(1-\alpha/2)\sigma/n^{\frac{1}{2}} .$$

For example, for a 95 percent confidence interval, we require the 0.975 quantile of the normal distribution, that is, the value of $z = z(0.975)$ such that

$$\Phi(z) = 0.975 .$$

From normal integral tables, we find $z = 1.96$. Hence, our confidence interval is $(\bar{x} - 1.96\sigma/n^{\frac{1}{2}}, \bar{x} + 1.96\sigma/n^{\frac{1}{2}})$ with coefficient 95 per cent.

Example. The software provided with most modern digital computers includes a library of sub-routines for a wide variety of applications. One such sub-routine which we shall use again later generates random numbers distributed as N(0,1). Using this procedure, we obtained the following 9 realizations of the random variable X.

+0.250, +1.620, -0.052, +0.017, -0.366, +0.756, +0.608, -2.150, +1.162.

We calculate the mean of this sample to be $\bar{x} = 0.205$. Hence, the 95 percent confidence interval for μ, the mean of the random number population, is

$$0.205 \pm 1.96 \times 1/3 = 0.205 \pm 0.653 ,$$

since the variance is 1. Of course, this example is somewhat artificial since we already know the true location of the population mean. Nevertheless, it is gratifying to observe that the interval calculated from a random sample drawn from this population and the given variance does in fact include the true value, zero.

Case 2. Any distribution with finite variance when n is large

We suppose that our observations, $X_1, X_2, \ldots X_n$, are independently, identically distributed with mean μ and variance σ^2, both unknown, and n is large. By the Central Limit Theorem, $\bar{X}$ is approximately distributed as $N(\mu, \sigma^2/n)$, so that $(\bar{X}-\mu)/(\sigma/n^{\frac{1}{2}}) \sim N(0,1)$. However, σ^2 is unknown, but we may make do with s^2 since it is also approximately true that $(\bar{X}-\mu)/(s/n^{\frac{1}{2}}) \sim N(0,1)$; here s^2 stands for the sample variance, that is

$$s^2 = \Sigma(X_i-\bar{X})^2/(n-1) \quad \text{or} \quad \Sigma(X_i-\bar{X})^2/n ,$$

since, with large n, the difference in the divisors is negligible. Therefore, we can treat this case in the same way as Case 1, just replacing σ by s. Thus, the confidence interval with coefficient $1-\alpha$ is

$$\bar{x} \pm z(1-\alpha/2)s/n^{\frac{1}{2}} ,$$

s now being the particular estimate calculated from the observations.

As a general guide, a sample size of at least 50 should qualify as a large sample. If we have some prior knowledge which suggests that the distribution of our observations resembles the normal in some way, e.g. symmetry, then this would allow us to regard smaller samples as large enough. On the other hand, strong deviations from normality, e.g. skewness, require more observations. In the absence of such prior information, inspection of the frequency distribution of the data may indicate similarity to the normal or otherwise and so help to resolve the question as to whether or not n can be regarded as sufficiently large.

Example. We use the data of Table 3.1 showing results obtained for the enthalpy of neutralization of NaOH with HCl. It will be shown in Chapter 8, Section 5 that the means of the different groups comprising these data are not significantly different. Hence, we may regard our 65 values of x as independent estimates of the same quantity, μ say, so that we may approximate the distribution of the grand mean $\bar{X}$ by $N(\mu, s^2/65)$ as explained above. From the data, we calculate

$$\bar{x} = 57.34 ,$$

$$s = 1.65 .$$

Now, if we wish to calculate a confidence interval with coefficient 0.90, we require the value of z such that $\Phi(z) = 0.95$. From tables, we obtain $z(0.95) = 1.64$. Hence, the 90 percent confidence interval for the distribution mean of our data is

$$(57.34 \pm 1.64 \text{x} 1.65/65^{\frac{1}{2}}) = 57.34 \pm 0.34 .$$

Case 3. Normal distribution with unknown variance

We have $X_1, X_2, \ldots X_n$ independently distributed as $N(\mu, \sigma^2)$ with both μ and σ^2 unknown, and we wish to obtain a confidence interval for μ. We use the result stated in Chapter 4, Section 10, that $T = (\bar{X}-\mu)/(s/n^{\frac{1}{2}})$ is distributed as t_{n-1}. Hence, T lies between $t_{n-1}(\alpha/2)$ and $t_{n-1}(1-\alpha/2)$ with probability $1-\alpha$, where $t_{n-1}(\alpha/2)$ and $t_{n-1}(1-\alpha/2)$ represent the $\alpha/2$ and $(1-\alpha/2)$ quantiles of the distribution of Student's t with n-1 degrees of freedom. Since the distribution of t is symmetrical, we have

$$t_{n-1}(\alpha/2) = -t_{n-1}(1-\alpha/2) \quad . \tag{2}$$

Hence

$$\left.\begin{aligned} 1-\alpha &= P\{t_{n-1}(\alpha/2) \leq T \leq t_{n-1}(1-\alpha/2)\} \\ &= P\{-t_{n-1}(1-\alpha/2)s/n^{\frac{1}{2}} \leq \bar{X}-\mu \leq + t_{n-1}(1-\alpha/2)s/n^{\frac{1}{2}}\} \\ &= P\{\bar{X} - t_{n-1}(1-\alpha/2)s/n^{\frac{1}{2}} \leq \mu \leq \bar{X} + t_{n-1}(1-\alpha/2)s/n^{\frac{1}{2}}\}, \end{aligned}\right\} \tag{3}$$

since the three pairs of inequalities in (3) are equivalent. Thus the confidence interval for μ with coefficient $1-\alpha$ is

$$\bar{x} \pm t_{n-1}(1-\alpha/2)s/n^{\frac{1}{2}} \quad .$$

The value of $t_{n-1}(1-\alpha/2)$ is obtained from a t table.

Example. The data of Table 5.4 provide a typical example. These data consist of five replicate observations of the density of a high polymer, H1. As explained in the previous chapter, we obtain

$$\bar{x} = 1.21510$$

and

$$s = 3.7 \times 10^{-4}$$

for the mean and standard deviation of this sample of size $n = 5$. We now assume that the data are random samples drawn from a normal population with mean μ and variance σ^2. In this case, we may calculate a 90 percent confidence interval using the results of this section and looking up the value of $t_{5-1}(1-0.05) = t_4(0.95) = 2.13$. Hence the 90 percent confidence interval is

$$1.21510 \pm 2.13 \times 3.7 \times 10^{-4}/5^{\frac{1}{2}} = 1.21510 \pm 0.00035 \quad .$$

1.1. Use of a t table - a digression

The t table is presented in different ways in different sets of statistical tables. Usually, the number of degrees of freedom is given down the side and, across the top the probability of t being greater than (or the probability of it being less than) the tabulated value. Alternatively, the top of the table may show the probability of $|t|$ being greater than (or the probability of it being less than) the tabulated value. Often these probabilities are expressed as percentages. The main difference is whether or not the probabilities relate to t or to $|t|$. The descriptive caption should explain which probabilities are meant, but one quick way to tell is by looking at the bottom line of the table (corresponding to ∞ degrees of freedom) under 5 percent (or 0.05); if the value is 1.96, the table relates to $|t|$; if it is 1.64, the table relates to t. These values, 1.96 and 1.64, correspond to the standard normal distribution (to which the distribution of t degenerates when the number of degrees of freedom becomes infinite) and should become familiar in time.

The degrees of freedom, ν, may sometimes need to be interpolated. Typically, the last few tabulated values of ν (in reverse order) are ∞, 120, 60, 40, 30, which are 120/0, 120/1, 120/2, 120/3, 120/4. To interpolate in this region, we use harmonic interpolation, taking the t value to be linear in $120/\nu$. To exemplify this, we take the case of $\nu = 49$ and the entries 2.704 ($\nu = 40$) and 2.660 ($\nu = 60$) at the probability level of 1 percent (two-tailed table). Since $120/49 \simeq 2.45$, the approximate critical value will be $2.660+0.45(2.704-2.660)$; the correction, in units of the third decimal place, is thus $0.45 \times 44 = 20$ so that the approximate critical value is 2.680.

2. Interval for difference of two means

We suppose we have two samples $X_1, X_2, \ldots X_{n_1}$ and $Y_1, Y_2, \ldots Y_{n_2}$, all the observations being independent. We write μ_1 and μ_2 for the distribution means of the X's and the Y's. Our object is to obtain a confidence interval for $\mu_1 - \mu_2$.

We consider four cases differing from one another in the information we have about the distribution of the X's and the Y's; the first three of these cases correspond to those of the last section.

Case 1. Normal distributions with known variances

Here we assume that the X's have the distribution $N(\mu_1,\sigma_1^2)$ and the Y's the distribution $N(\mu_2,\sigma_2^2)$, σ_1^2 and σ_2^2 both being known. Our estimator for $\mu_1-\mu_2$ is $\bar{X}-\bar{Y}$, the variance of which is $\sigma_1^2/n_1 + \sigma_2^2/n_2$; furthermore,

$$(\bar{X}-\bar{Y}) - (\mu_1-\mu_2) \sim N(0,\sigma_1^2/n_1 + \sigma_2^2/n_2) ,$$

so that the quantity T given by

$$T = \frac{(\bar{X}-\bar{Y}) - (\mu_1-\mu_2)}{(\sigma_1^2/n_1 + \sigma_2^2/n_2)^{\frac{1}{2}}}$$

is distributed as N(0,1). If $z(1-\alpha/2)$ is such that $\Phi\{z(1-\alpha/2)\} = 1-\alpha/2$, then

$$\begin{aligned} P\{|T| \leq z(1-\alpha/2)\} &= 1 - P\{|T| > z(1-\alpha/2)\} \\ &= 1 - 2[1-\Phi\{z(1-\alpha/2)\}] \\ &= 1 - 2(\alpha/2) \\ &= 1 - \alpha . \end{aligned}$$

Hence

$$P\{|(\bar{X}-\bar{Y}) - (\mu_1-\mu_2)| \leq (\sigma_1^2/n_1 + \sigma_2^2/n_2)^{\frac{1}{2}} z(1-\alpha/2)\} = 1 - \alpha . \quad (1)$$

This equation shows that a confidence interval for $\mu_1-\mu_2$ with coefficient $1-\alpha$ is

$$(\bar{x}-\bar{y}) \pm (\sigma_1^2/n_1 + \sigma_2^2/n_2)^{\frac{1}{2}} z(1-\alpha/2) .$$

This case is rather unusual and so we shall not pursue it further.

Case 2. Any distributions with finite variances when n_1 and n_2 are large

Here the variances of the X's and the Y's, σ_1^2 and σ_2^2, are unknown.

Proceeding analogously to Case 2 of the previous section, we simply replace σ_1^2 and σ_2^2 throughout the above derivation by the sample variances s_1^2 and s_2^2. Thus, the confidence interval for $\mu_1 - \mu_2$ with coefficient $1-\alpha$ is

$$(\bar{x}-\bar{y}) \pm (s_1^2/n_1 + s_2^2/n_2)^{\frac{1}{2}} z(1-\alpha/2) \ .$$

The sample variances may be evaluated using either n or n-1 as the divisor since the difference is negligible when n is large; that is

$$s_1^2 = \Sigma(x_i-\bar{x})^2/(n_1-1) \quad \text{or} \quad \Sigma(x_i-\bar{x})^2/n_1 \ ,$$

and

$$s_2^2 = \Sigma(y_i-\bar{y})^2/(n_2-1) \quad \text{or} \quad \Sigma(y_i-\bar{y})^2/n_2 \ .$$

Example. The enthalpy of neutralization data shown in Table 3.1 were obtained by a class of students in 1968. In addition to these data, we also have available similar results obtained by the class of 1967. The results for this class, which unlike the class of the following year were not split into groups, are shown in Table 6.1.

Table 6.1. Values of the enthalpy change accompanying the neutralization of NaOH using HCl

54.4	56.9	58.4	56.1	56.5
56.4	57.5	57.0	56.8	58.0
57.5	56.4	55.2	56.4	57.2
56.6	54.4	58.2	57.4	57.7
57.0	59.0	57.2	56.3	57.9
56.5	57.9	57.8	56.4	58.0
58.3	56.5			

The above are the numerical values of $-\Delta H/(\text{kJ mol}^{-1})$

It is interesting to compute a confidence interval for the difference between the distribution means of the two classes. The relevant data are assembled below, using X and Y to distinguish the observations of the 1968 and 1967 classes.

$$\bar{x} = 57.34 \qquad \bar{y} = 56.99$$

$$s^2(x) = 2.726 \qquad s^2(y) = 1.130$$

$$n_1 = 65 \qquad n_2 = 32$$

Here $\bar{x}$ and $\bar{y}$ stand for the grand means of the values of $-\Delta H/(\text{kJ mol}^{-1})$ for the two classes. Referring to Section 1, Case 2 and noting that we expect our data to be approximately symmetrically distributed, we may regard the sample sizes of 65 and 32 as large. From the previous arguments, we compute that the confidence interval for $\mu_{1968} - \mu_{1967}$ with coefficient 0.90 is

$$(57.34 - 56.99) \pm (2.726/65 + 1.130/32)^{\frac{1}{2}} z(0.95) = +0.35 \pm 0.46 \ .$$

We note that the interval includes zero (see also Chapter 8, Section 2, Case 2).

<u>Case 3</u>. Normal distributions with the same but unknown variance

Here the X's are distributed as $N(\mu_1,\sigma^2)$, the Y's as $N(\mu_2,\sigma^2)$, and σ^2, assumed the same for each distribution, is unknown. If n_1 and n_2 were both large, we could obtain an interval as in Case 2. However, we are not assuming that n_1 and n_2 are large in the present case.

Under these circumstances, we have already seen in Chapter 4 that the quantity T given by

$$T = \frac{(\bar{X}-\bar{Y}) - (\mu_1-\mu_2)}{s\{1/n_1 + 1/n_2\}^{\frac{1}{2}}}$$

is distributed as $t_{n_1+n_2-2}$, where s, the pooled estimator of

σ, is given by

$$s^2 = \frac{\Sigma(X_i-\bar{X})^2 + \Sigma(Y_i-\bar{Y})^2}{n_1 + n_2 - 2} . \qquad (2)$$

Then

$$P\{|T| \le t_{n_1+n_2-2}(1-\alpha/2)\} = 1-\alpha .$$

That is,

$$P\{|(\bar{X}-\bar{Y}) - (\mu_1-\mu_2)| \le s(1/n_1+1/n_2)^{\frac{1}{2}} t_{n_1+n_2-2}(1-\alpha/2)\} = 1-\alpha , \qquad (3)$$

where $t_{n_1+n_2-2}(1-\alpha/2)$ is the $1-\alpha/2$ quantile of the distribution of t with n_1+n_2-2 degrees of freedom.

Equation (3) shows that the appropriate confidence interval for $\mu_1-\mu_2$ with coefficient $1-\alpha$ is

$$(\bar{x}-\bar{y}) \pm s(1/n_1 + 1/n_2)^{\frac{1}{2}} t_{n_1+n_2-2}(1-\alpha/2) ,$$

where s now stands for the realization of the right-hand side of equation (2).

Example. The data of Table 6.2 taken together with those of Table 5.4 provide a useful illustration of the application of the above procedure. These new data refer to a second high polymer, H2, prepared under ostensibly the same conditions as H1, but using reagents drawn from new stock. The numerical values in the table are the densities of five separate pieces of H2; we shall label these data $Y_1,Y_2,\ldots Y_5$ and the H1 set $X_1,X_2,\ldots X_5$.

Table 6.2. Densities of five samples of a high molecular weight polymer, H2

Sample	1	2	3	4	5
Density/g cm^{-3}	1.2167	1.2176	1.2157	1.2158	1.2167

We assume that each set are random samples drawn from two normal populations with different μ but the same σ^2. We have no real evidence for the assumption of normality (the sample size is too small), but the values of the two sample variances, viz.

$$s^2(x) = 14.0 \times 10^{-8}$$

$$s^2(y) = 60.5 \times 10^{-8}$$

suggest that the assumption of the same variance is not unreasonable (we shall see in Chapter 9 how to confirm this). Hence the pooled estimate of σ^2 is

$$\left(\frac{56+242}{5+5-2}\right) \times 10^{-8} = 37.25 \times 10^{-8} .$$

We now assemble together the various numerical values required to calculate a 90 percent confidence interval for the difference $\mu_{H1} - \mu_{H2}$. These are

$$\bar{x} = 1.21510 ,$$

$$\bar{y} = 1.21650 ,$$

$$n_1 = 5 ,$$

$$n_2 = 5 ,$$

$$t_8(0.95) = 1.86 .$$

Hence the confidence interval is

$$(1.21510-1.21650) \pm 6.10 \times 10^{-4} \times (2/5)^{\frac{1}{2}} \times 1.86 = -0.00140 \pm 0.00072 .$$

We note that the interval does not include zero.

Case 4. Normal distributions with unequal unknown variances

We assume here that the X's are distributed as $N(\mu_1,\sigma_1^2)$, the Y's as $N(\mu_2,\sigma_2^2)$ so that this case is the same as Case 3 except that $\sigma_1^2 \neq \sigma_2^2$. This situation is more difficult to handle than the previous case and presents what is known as the Fisher-Behrens problem. Actually, we may use the procedure of Case 3 as a reasonable approximation if σ_1^2 and σ_2^2 are not too different, especially if $n_1 = n_2$.

More generally, an approximate solution due to Welch and Aspin (1967, 1968, 1969) will usually be more satisfactory. We use the random variable

$$T = \frac{(\bar{X}-\bar{Y}) - (\mu_1-\mu_2)}{(s_1^2/n_1 + s_2^2/n_2)^{\frac{1}{2}}} ,$$

where $s_1^2 = \Sigma(X_i-\bar{X})^2/(n_1-1)$ and $s_2^2 = \Sigma(Y_i-\bar{Y})^2/(n_2-1)$. Now T is approximately distributed as t with ν degrees of freedom, where ν is given by

$$\frac{1}{\nu} = \frac{1}{\nu_1}\left(\frac{s_1^2/n_1}{s_1^2/n_1 + s_2^2/n_2}\right)^2 + \frac{1}{\nu_2}\left(\frac{s_2^2/n_2}{s_1^2/n_1 + s_2^2/n_2}\right)^2 . \quad (4)$$

In this expression ν_1 and ν_2 stand for (n_1-1) and (n_2-1), the numbers of degrees of freedom associated with s_1^2 and s_2^2 respectively. We see that ν is a weighted average of ν_1 and ν_2.

To obtain the confidence interval for $(\mu_1-\mu_2)$, we simply adapt the arguments of Case 3 to show that the interval with coefficient $1-\alpha$ is

$$(\bar{x}-\bar{y}) \pm (s_1^2/n_1 + s_2^2/n_2)^{\frac{1}{2}}\, t_\nu(1-\alpha/2) .$$

Example. In a study of the relative merits of two different methods for the determination of the amount of a substance W in solution, ten replicate measurements were made using each method. The results are shown in Table 6.3. In the first instance, we look for a systematic difference between the two methods and to this end we calculate a confidence interval for $\mu_A - \mu_B$. A coefficient of 0.90 is appropriate here. From the data in the table, we obtain

method A	method B
$\bar{x} = 5.140$	$\bar{y} = 5.203$
$s_1^2 = 11.11 \times 10^{-4}$	$s_2^2 = 45.34 \times 10^{-4}$
$n_1 = 10$	$n_2 = 10$

Table 6.3. The percentage of component W in a solution as determined by two different analytical methods

Determination	Method A	Method B
1	5.11	5.18
2	5.14	5.13
3	5.13	5.27
4	5.17	5.12
5	5.12	5.27
6	5.08	5.29
7	5.15	5.17
8	5.20	5.28
9	5.16	5.14
10	5.14	5.18

The ratio s_2^2/s_1^2 can be shown by the methods of Chapter 9 to be such that we should not regard the variances as equal. Using equation (4) we calculate ν; thus

$$\frac{1}{\nu} = \frac{1}{9}\left(\frac{1.111}{5.645}\right)^2 + \frac{1}{9}\left(\frac{4.534}{5.645}\right)^2$$

$$\nu = 13.2 .$$

From t tables, we find $t_{13.2}(0.95) \simeq 1.77$. Hence, the confidence interval for $\mu_A - \mu_B$ is

$$-0.063 \pm (1.111 + 4.534)^{\frac{1}{2}} \times 10^{-2} \times 1.77$$

$$= -0.063 \pm 0.042 .$$

This result suggests that there is a systematic difference between the results obtained by method A and those obtained by method B, since the interval does not include zero.

3. Intervals from large samples of binomially distributed data

We consider here a large sample of data which are distributed as B(p,n). As we have seen in Chapter 4, if the event which we label 'success' has a probability of occurrence of p in each independent performance of a binomial trial, then the number of successes from n trials has the B(p,n) distribution. If n is large, the proportion of successes in the sample, $\hat{p}$ (the maximum likelihood estimator of p) is approximately distributed as N(p,pq/n) by the Central Limit Theorem, q being simply 1-p. Hence the quantity

$$T_1 = \frac{\hat{p} - p}{(pq/n)^{\frac{1}{2}}}$$

is distributed as N(0,1) approximately. It is also approximately true that

$$T_2 = \frac{\hat{p} - p}{(\hat{p}\hat{q}/n)^{\frac{1}{2}}} ,$$

where $\hat{q} = 1-\hat{p}$, is distributed as N(0,1) although this approximation is not quite as good as that for T_1.

3.1. Interval for p

The interval for p can now be obtained just as in Case 2 of Section 1 merely using $\hat{p}\hat{q}/n$ as the variance estimator. We have

$$P\{\left| \frac{\hat{p} - p}{(\hat{p}\hat{q}/n)^{\frac{1}{2}}} \right| \leq z(1-\alpha/2)\} = 1 - \alpha , \tag{1}$$

so that the confidence interval for p with coefficient 1-α is

$$\hat{p} \pm (\hat{p}\hat{q}/n)^{\frac{1}{2}} z(1-\alpha/2) .$$

A more accurate confidence interval for p can be obtained using the fact that T_1 above is closer to a standard normal variate than T_2. Consequently, with greater validity, we may write in place of equation (1)

$$P\{|\frac{\hat{p} - p}{(pq/n)^{\frac{1}{2}}}| \le z(1-\alpha/2)\} = 1 - \alpha . \tag{2}$$

Squaring both sides of the inequality in equation (2) and replacing q by 1-p gives

$$P\{(\hat{p} - p)^2 \le p(1-p)z(1-\alpha/2)/n\} = 1 - \alpha .$$

The inequality inside the bracket holds between the two roots of the quadratic

$$(\hat{p} - p)^2 - p(1-p)z(1-\alpha/2)/n = 0 . \tag{3}$$

Hence, the roots of equation (3) represent the confidence interval for p with coefficient $1-\alpha$.

Example. In an objective examination, 40 multiple choice questions are set. Each question is followed by four possible responses only one of which is the correct answer to the question. A particular candidate selects 12 correct answers out of the 40 and we wish to calculate a 90 percent confidence interval for p, the probability of his getting a random question in the examination correct.

Using the method based on the approximate normality of T_2 and choosing a confidence coefficient of 0.90, we obtain

$$\hat{p} = 0.30 ,$$

$$\hat{q} = 0.70 ,$$

$$z(0.95) = 1.64 ,$$

so that our confidence interval for p is (0.18,0.42). If the method based on T_1 is used instead, the interval becomes (0.22,0.40). We notice that both these intervals include the value 0.25, which is the value for complete ignorance when each answer is randomly selected. Neither interval gives any evidence against the hypothesis of complete ignorance.

We should draw attention to the way in which the interval widens for a given $\hat{p}$ as the number of trials decreases. That is, the fewer trials there are, the more uncertain is the location of the true value of the parameter, p. Thus if there had been only 20 questions instead of 40 and if our candidate had answered 6 correctly so that $\hat{p} = 0.30$ as before, the interval for p with coefficient 0.90 would have been (0.13,0.47), based on the quantity T_2. If these arguments are continued, it is clear that the categorization of candidates is much less certain with a smaller number of questions.

3.2. Interval for difference of two values of p

Here, we suppose we have two independent binomial trials, one producing data distributed as $B(p_1,n_1)$ and the other producing data distributed as $B(p_2,n_2)$. We imagine that we sample these two distributions in two separate experiments and, from the proportions of successes in the two cases, estimate p_1 by $\hat{p}_1$ and p_2 by $\hat{p}_2$. Proceeding as in Case 2 of Section 2 and using $\hat{p}_1\hat{q}_1/n_1$ and $\hat{p}_2\hat{q}_2/n_2$ as the two variance estimates, we obtain

$$(\hat{p}_1 - \hat{p}_2) \pm (\hat{p}_1\hat{q}_1/n_1 + \hat{p}_2\hat{q}_2/n_2)^{\frac{1}{2}}z(1-\alpha/2)$$

for the confidence interval for p_1-p_2 with coefficient $1-\alpha$.

Example. An interesting example is provided by the regional results of the voting in the British Referendum in 1975 as to whether or not Britain should remain in the EEC. Let us suppose each voter in Scotland has a probability of p_1 of voting 'yes'. Then the proportion of 'yes' votes recorded for Scotland gives us an estimate $\hat{p}$ of p_1 from n_1 independent trials, where n_1 is the total number of 'yes' and 'no' votes. We may use a similar description for the results for Northern Ireland, simply replacing the subscript 1 by 2. We require a confidence interval for p_1-p_2 with coefficient 0.90. First we assemble the relevant data.

	Yes	1332186	$\hat{p}_1$	0.58430
Scotland	No	947769	$\hat{q}_1$	0.41570
	n_1	2279955	$\hat{p}_1\hat{q}_1/n_1$	1.065×10^{-7}

N. Ireland	Yes	259251	$\hat{p}_2$	0.52146
	No	237911	$\hat{q}_2$	0.47854
	n_2	497162	$\hat{p}_2\hat{q}_2/n_2$	5.019×10^{-7}

Hence the confidence interval with coefficient 0.90 for p_1-p_2 is

$$(0.58430 - 0.52146) \pm (10.65 + 50.19)^{\frac{1}{2}} \times 10^{-4} \times 1.64$$

$$= 0.0628 \pm 0.0013.$$

We note that the interval does not include zero.

4. Interval for a variance when the observations are normally distributed

We suppose that we have observations $X_1, X_2, \ldots X_n$ independently distributed as $N(\mu,\sigma^2)$, where both μ and σ^2 are unknown. Then as we saw in Chapter 4, Section 10, $\Sigma(X_i-\bar{X})^2/\sigma^2$ is distributed as χ^2_{n-1}. Therefore we may write

$$P\{\chi^2_{n-1}(\alpha/2) \leq \Sigma(X_i-\bar{X})^2/\sigma^2 \leq \chi^2_{n-1}(1-\alpha/2)\} = 1 - \alpha ,$$

where $\chi^2_{n-1}(\alpha/2)$ and $\chi^2_{n-1}(1-\alpha/2)$ represent the $\alpha/2$ and $(1-\alpha/2)$ quantiles of the distribution of χ^2_{n-1}. Hence

$$P\left\{\frac{\Sigma(X_i-\bar{X})^2}{\chi^2_{n-1}(1-\alpha/2)} \leq \sigma^2 \leq \frac{\Sigma(X_i-\bar{X})^2}{\chi^2_{n-1}(\alpha/2)}\right\} = 1 - \alpha .$$

Hence the confidence interval for σ^2 with coefficient $1-\alpha$ is

$$\left(\frac{\Sigma(x_i-\bar{x})^2}{\chi^2_{n-1}(1-\alpha/2)} , \frac{\Sigma(x_i-\bar{x})^2}{\chi^2_{n-1}(\alpha/2)}\right) .$$

Alternatively, since $\Sigma(x_i-\bar{x})^2 = (n-1)s^2$, the interval may be written

$$\left(\frac{(n-1)s^2}{\chi^2_{n-1}(1-\alpha/2)} , \frac{(n-1)s^2}{\chi^2_{n-1}(\alpha/2)}\right) .$$

Note that it is not possible to write this in the form $s^2 \pm e$ since the distribution of χ^2 is not symmetrical.

Example. We return to our density data on the polymer H1. We have seen in Chapter 5, Section 4 that s^2 for these data is 14.0×10^{-8} based on 4 degrees of freedom. If we require a 90 percent confidence interval for σ^2, we need the values of $\chi_4^2(0.05)$ and $\chi_4^2(0.95)$. From tables, we find these to be 0.711 and 9.49 respectively. Hence the required confidence interval is $(5.90, 78.8) \times 10^{-8}$. The large width of the interval shows how poorly located is the value of σ^2 when only a limited number of degrees of freedom is available for its estimation.

5. Approximate interval for a function

Here, we use the approximate mean and variance of a function as given in Chapter 5, Section 5. We assume that the estimating expression is approximately normal, an assumption which would apply, for example, if the estimated arguments are maximum likelihood estimates of the true arguments based on a large number of observations. Another case where this assumption of approximate normality may sometimes be used is when the errors in the arguments (of which there should not be too few) are due to independent, uniformly-distributed rounding-off error as illustrated in Chapter 5, Section 6.

Under these assumptions, the approximate interval with coefficient $1-\alpha$ is given by the procedure of Section 1, Case 2 when the T's are independent, as

$$\phi(t_1, t_2, \ldots t_r) \pm z(1-\alpha/2)\left\{\sum\left(\frac{\partial\phi}{\partial\theta_i}\right)^2 s^2(T_i)\right\}^{\frac{1}{2}} .$$

Here t_i is the realization of the estimator T_i of θ_i, $i = 1,2,\ldots r$; the derivatives $\{\partial\phi/\partial\theta_i\}$ are evaluated at the values of $\{t_i\}$.

An example of this procedure has already been given in Chapter 5, Section 6, where we calculated the stochastic error due to rounding off.

CHAPTER 7

HYPOTHESIS TESTING

In the preceding chapter, we have explained how we may obtain some idea of the location of the parameters of a distribution from the corresponding estimates and a confidence interval. Often this information enables us to draw further inferences from the data which may be of greater importance than the location itself. For example, referring to the results of Section 3.2 of the previous chapter, it does seem as if there is a real difference between the preferences of the voters in Scotland and Northern Ireland for continued British membership of the EEC. Although we can see this in an intuitive way, there is clearly an advantage in having a formal procedure for coming to this conclusion. This chapter deals in a quite general way with such procedures.

1. Null and alternative hypotheses

In most cases we suppose that our data come from a distribution or distributions of known form, but with one or more unknown parameters. Typically, these parameters will be the mean and/or variance. We set up two alternative hypotheses about these parameters which we shall represent as H_0 and H_1 and we shall refer to these as the null and alternative hypotheses respectively. The null hypothesis is the hypothesis under test, or the hypothesis being challenged. Often H_0 will state that some parameter or some difference between parameters is zero, which links with the word null. It is important to emphasize that we regard these two hypotheses as mutually exclusive and exhaustive. Thus, if, for some reason, we decide to accept H_0, then we reject H_1; conversely rejection of H_0 means acceptance of H_1. This is not to say that a particular formulation of H_1 is the one and only alternative to H_0. The situation may be that it is possible to conceive of several plausible alternatives to H_0, say H_1', H_1'' and so on. However, once H_0 and H_1 have been formulated, then the reasoning leading to the acceptance/rejection of H_0

necessarily leads to the rejection/acceptance of H_1. Clearly it is the responsibility of the experimentalist to select the most appropriate pair of hypotheses for his particular situation.

We distinguish two general types of hypothesis which we call <u>simple</u> and <u>compound</u> according to whether or not the hypothesis fully specifies the parameter(s) of the distribution. A typical simple null hypothesis might be

$$H_0: \theta = \theta_0 ,$$

that is, the unknown parameter θ of the distribution from which our data are drawn is equal to some particular value, θ_0. Usually, the alternative hypothesis is compound, since it is often enough for H_1 to assert that θ is not located at the value θ_0 without specifying it further; thus three possible choices for H_1 which pair with the above H_0 and which are compound are:

$$H_1 : \theta > \theta_0 , \quad \text{or} \quad H_1 : \theta < \theta_0 , \quad \text{or} \quad H_1 : \theta \neq \theta_0 .$$

In the first two of these H_1 is <u>one-sided</u>, in the last it is <u>two-sided</u>. It will sometimes be convenient for us to express our alternative hypothesis, in a general form, as 'not H_0', meaning simply the complement of H_0. When the set of possible θ values is restricted as, for example, in $H_0 : \theta = \theta_0$ with $H_1 : \theta > \theta_0$, 'not H_0' is this H_1 not $\theta \neq \theta_0$. It is also possible to have H_0 compound. For example, the first two formulations of H_1 may be associated with the corresponding null hypotheses

$$H_0 : \theta \geq \theta_0 , \quad \text{or} \quad H_0 : \theta \leq \theta_0 .$$

2. <u>Tests</u>

We may decide whether to accept or reject H_0 according as an estimate θ^* of θ computed from the data is too large, or too small compared with the null hypothetical value θ_0. For this purpose, we have to decide upon a test which is <u>one-sided</u> or <u>two-sided</u> according as H_1 is one- or two-sided. The basic situations which may occur are thus:

(a) <u>one-sided</u>

$H_0 : \theta = \theta_0$ (or $H_0 : \theta \leq \theta_0$)
$H_1 : \theta > \theta_0$
} We reject H_0 if θ^* is too <u>large</u> compared with θ_0.

or

$H_0 : \theta = \theta_0$ (or $H_0 : \theta \geq \theta_0$)
$H_1 : \theta < \theta_0$
} We reject H_0 if θ^* is too <u>small</u> compared with θ_0.

(b) <u>two-sided</u>

$H_0 : \theta = \theta_0$
$H_1 : \theta \neq \theta_0$
} We reject H_0 if θ^* is too <u>different</u> from θ_0, above or below it.

In general, we may expect to adopt a two-sided test unless we have some <u>prior</u> knowledge that θ, if not equal to θ_0, can only lie to one side of it, because of some systematic effect in the measurements. It is invalid to use the data themselves as the basis for deciding whether or not to perform a one- or two-sided test. The two-sided test is always the more severe, requiring stronger evidence for the rejection of H_0 than for its rejection in a corresponding one-sided test. When we have rejected a simple H_0 in favour of a two-sided H_1, strictly we can then only accept θ as being <u>different</u> from θ_0. However, it is natural and reasonable to suppose that θ lies on the same side of θ_0 as θ^* does. Certainly, it would be reasonable for us to use θ^* as our estimate of θ, as discussed in Section 5.

In order to decide whether θ^* is too large, too small, or too different from θ_0, we use some appropriate test statistic T, rejecting H_0 if T is either too large or too small as the case may be. Usually, it is impossible to devise a test which will be infallible, always rejecting H_0 if it is false and accepting it if it is true. We have to work with 'calculated risks' so to speak.

Hence when we state the conclusion of our test, we say

that we <u>accept</u> H_0 (or H_1, as the case may be). We <u>do not</u> say, "therefore H_0 (say) is true", or "we have proved the truth of H_0", or such statements as these. Rather we <u>assume</u> the truth of H_0, <u>behave as if</u> it were true, in our future work.

The generally accepted (Neyman-Pearson) theory of hypothesis testing protects H_0 at the expense of the alternative hypothesis, H_1. This is appropriate since usually H_0 is the hypothesis of 'no effect' or 'no change', calling for accepting the status quo and not effecting any modifications, whereas acceptance of H_1 usually requires action to be taken, practical changes to be made. Hence, it is reasonable to protect the status quo, the previously accepted beliefs, and to require any innovations indicated by H_1 to be firmly supported by sufficient evidence.

The most common procedure is to fix a <u>level of significance</u>, α, sometimes called <u>size</u>, of our test, which is the <u>maximum</u> probability of rejecting H_0 when H_0 is true. The significance levels which are usually used are 5, 1 and 0.1 percent in descending order of popularity. Often α is $P(T \geq c|H_0)$, the probability that our test statistic, T, is greater than or equal to some value c, called the <u>critical</u> value, given that H_0 applies. We then reject H_0 at the significance level α if a realization, t, of our test statistic is such that $t \geq c$. In other cases, α may be equal to $P(|T| \geq c|H_0)$ and so we reject H_0 when $|t| \geq c$.

When we are comparing a particular value t (or $|t|$) with a critical value, c, at a certain significance level, α, and t (or $|t|$) is 'close to' c, we might hesitate at using the strict rule of rejecting H_0 if and only if t (or $|t|$) $\geq c$. If the issue is important, more data should be obtained.

Alternatively rather than saying that T is significantly large at a certain level and rejecting H_0 on this basis, we may state more informatively the probability under H_0 of T being greater than or equal to the value, t, actually calculated in the test, which is called the <u>significance probability</u> and which we shall represent in this chapter by P.

Before going any further, we illustrate diagrammatically the meaning of significance level for the two hypotheses

$$H_0 : \theta = \theta_0 \quad ,$$

$$H_1 : \theta > \theta_0 \quad ,$$

when our data comprise n independent observations, X, drawn from a normal population of known variance, σ^2. Now

$$X \sim N(\theta,\sigma^2) \quad ,$$

$$\bar{X} \sim N(\theta,\sigma^2/n) \quad .$$

Hence, a test statistic, T, defined by

$$T = \frac{\bar{X} - \theta_0}{\sigma/n^{\frac{1}{2}}} \quad ,$$

is distributed as N(0,1) under H_0 and as $N(\eta,1)$ under H_1, where $\eta = (\theta-\theta_0)/(\sigma/n^{\frac{1}{2}})$. The density curves for these two cases are shown in Figure 7.1. Suppose we fix a level of significance α, which is such that

$$P(T \geq c|H_0) = \alpha$$

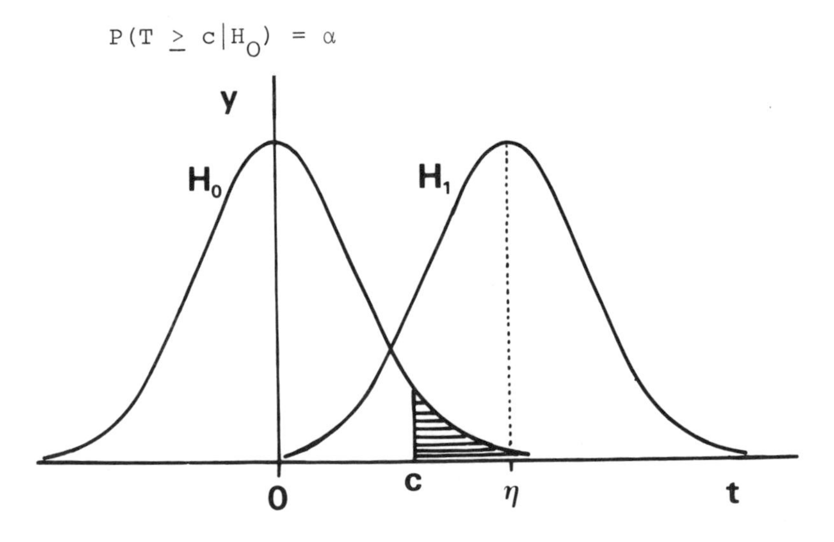

Figure 7.1. Illustration of significance level. Graphs of $y = f(t|H)$ for H_0 and H_1. The shaded area under the curve labelled H_0 represents a fraction α

We have marked c on Figure 7.1. Then if t falls into the shaded area, we reject H_0 knowing full well that we shall be incorrect in doing so in a proportion α of tests if H_0 is true. Alternatively we may calculate

$$P(T \geq t|H_0) = P \quad ,$$

the significance probability which is also illustrated by Figure 7.1 with c replaced by t. We may thus reject H_0 at any significance level greater than or equal to P.

3. Type I and type II errors

Rejection of H_0 when H_0 is true (or acceptance of H_1 when H_1 is false) is called a type I error and we shall call its probability P(I). If H_0 is simple, then P(I) is equal to α. If H_0 is compound, P(I) will be a function of the not-fully-specified parameter θ (which may be a vector) and $\alpha = \max P(I)$ under H_0. To make this last point clearer, we modify our previous example slightly, changing the hypotheses to

$$H_0 : \theta \leq \theta_0 \quad ,$$
$$H_1 : \theta > \theta_0 \quad .$$

Then for n normally distributed observations, $X_1, X_2, \ldots X_n$, the test statistic $T = (\bar{X}-\theta_0)/(\sigma/n^{\frac{1}{2}})$ is such that $T \sim N(\eta,1)$. From Figure 7.2, we see clearly that, as the true value of θ decreases from θ_0, the area of the tail of the distribution lying above c decreases. This means that the probability of our rejecting H_0 falsely decreases with decreasing θ and that the maximum value of P(I) is α, the area under the tail of the distribution corresponding to $\theta = \theta_0$. Figure 7.3 shows P(I) plotted against θ for this case.

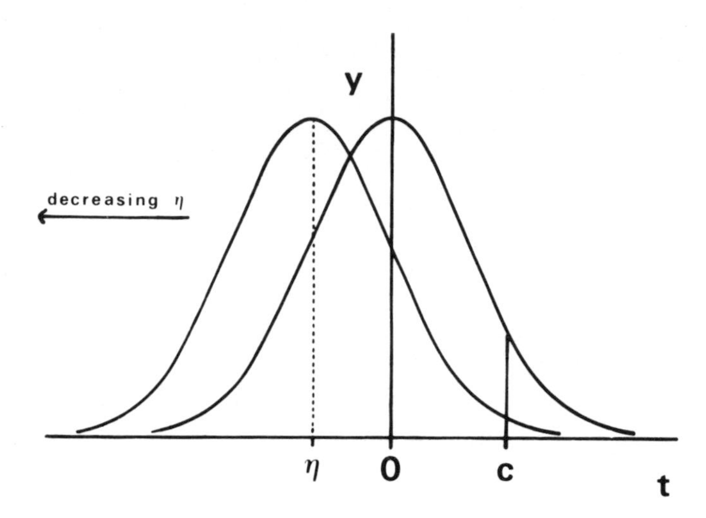

Figure 7.2. Illustration of the variation of P(I) with η (or θ) for compound $H_0 : \theta \leq \theta_0$. Graphs of $y = f(t|H_0)$ for $\eta = 0$ and $\eta < 0$. The areas of the tails of the curves beyond c represent P(I), c being fixed by the $\eta = 0$ case, viz.

$$P\{T = (\bar{X} - \theta_0)/(\sigma/n^{\frac{1}{2}}) \geq c|H_0\} = \alpha .$$

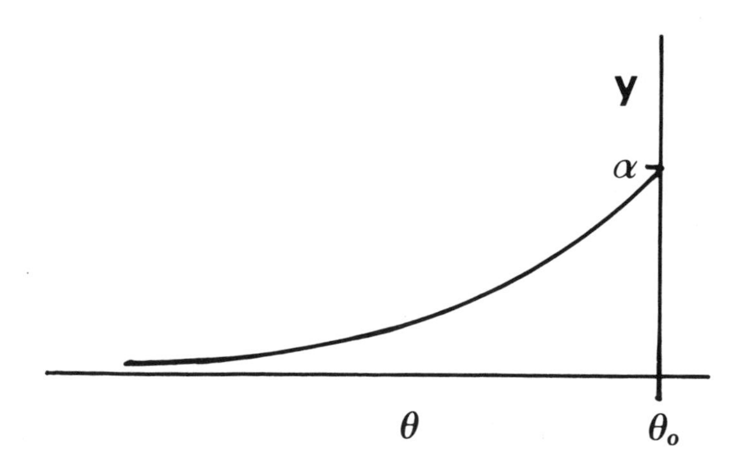

Figure 7.3. $y = P(I)$ as a function of θ for $H_0 : \theta \leq \theta_0$

A type II error is that of accepting H_0 when it is false (or rejecting H_1 when it is true) and its probability $P(II)$ or β will usually be a function of θ since H_1 is usually compound. To explain this, we use the same two curves as Figure 7.1 displayed for the present purpose in Figure 7.4. These two curves show the density functions for the statistic $T = (\bar{X}-\theta_0)/(\sigma/n^{\frac{1}{2}})$ under the two hypotheses $H_0 : \theta = \theta_0$ and $H_1 : \theta > \theta_0$ for n independent, normally distributed observations of X. Now suppose we use the same criterion as before, accepting H_0 if $t < c$ where c is such that $P(T \geq c|H_0) = \alpha$. If H_1 were true, we see that we should observe values of t to the left of c in proportion to the shaded area under the tail of the distribution corresponding to H_1. This shaded area, therefore, gives the probability of accepting H_0 falsely. We note that this area will decrease as θ increases from θ_0 showing that β is a function of θ, say $\beta(\theta)$; conversely as θ decreases towards θ_0, so the probability of a type II error increases until, when θ is just above θ_0, β is just below $1-\alpha$.

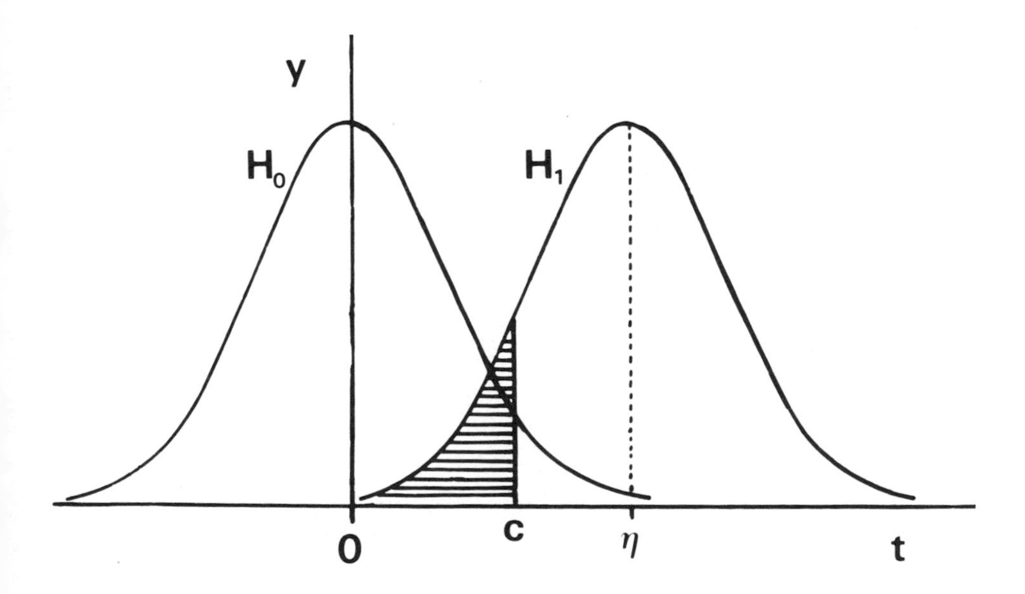

Figure 7.4. Illustration of $\beta = P(II)$. Graphs of $y = f(t|H)$ for H_0 and H_1. The shaded area under the curve labelled H_1 represents a fraction β.

It is useful to consider the probability of rejecting H_0 as a function of θ. Since H_0 is either true or false, there are clearly two quantities of interest, viz.

$$P(\text{rejecting } H_0|H_0 \text{ true}) \text{ and } P(\text{rejecting } H_0|H_0 \text{ false}) .$$

Now

$$\begin{aligned} P(\text{rejecting } H_0|H_0 \text{ true}) &= P(I) \\ &= \alpha \text{ when } H_0 \text{ is simple} \\ &\leq \alpha \text{ when } H_0 \text{ is compound.} \end{aligned}$$

Also

$$\begin{aligned} P(\text{rejecting } H_0|H_0 \text{ false}) &= 1 - P(\text{accepting } H_0|H_0 \text{ false}) \\ &= 1 - \beta(\theta) . \end{aligned}$$

This latter probability is called the power function, $\pi(\theta)$, of the test. Clearly it is desirable to have the power, $\pi(\theta)$, for any given θ, as large as possible. This depends upon the choice of an efficient test statistic, T. Furthermore, it is desirable to have $\pi(\theta) > \alpha$ for all θ under H_1, since this means that there is a greater probability of rejecting H_0 when it is false than rejecting it when it is true. When this condition is fulfilled, the test is called unbiased. Most of the tests which we shall give in the next few chapters use efficient test statistics and are unbiased, but we shall not discuss these aspects further.

Figure 7.5 shows the various quantities discussed above displayed diagrammatically for a one-sided situation. Similar curves may be drawn for the other situations.

It should be noted that we have discussed only the probability of rejecting H_0 given that it is true or given that it is false. We cannot say what the probability is that H_0 is true or that H_0 is false. This distinction is important since it is all too easy to slip from the statement that H_0 is accepted at the 5 percent significance level to the incorrect statement that the probability of H_0 being true is 95 percent.

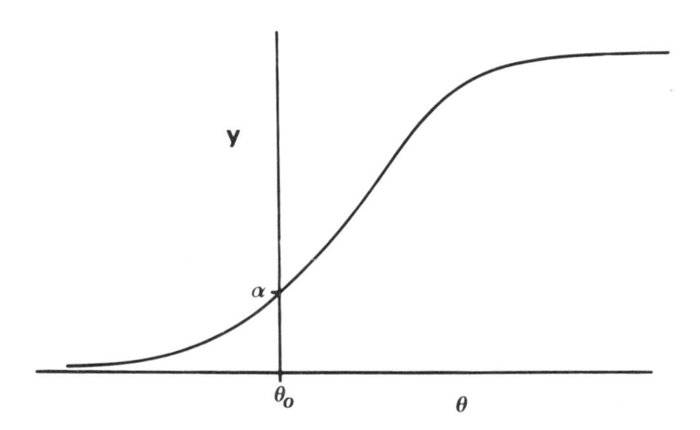

Figure 7.5 $y = P(\text{rejecting } H_0)$ as a function of θ for $H_0 : \theta \leq \theta_0$; $H_1 : \theta > \theta_0$. To the left of the line $\theta = \theta_0$, the ordinate represents P(I). To the right of this line, the ordinate represents $\pi(\theta) = 1 - P(II)$. Note that this figure is an extension of Figure 7.3. If the null hypothesis had been $H_0 : \theta = \theta_0$, the portion to the left of the line $\theta = \theta_0$ would be missing.

4. Correspondence with confidence interval

There exists a correspondence between a confidence interval for a parameter θ and a corresponding two-sided test of a simple null hypothesis:

$$H_0 : \theta = \theta_0 ,$$

$$H_1 : \theta \neq \theta_0 .$$

We can derive a test with significance level α from a confidence interval for θ with coefficient 1-α by deciding to accept H_0 if θ_0 lies in the interval, rejecting H_0 otherwise. Similarly, we can obtain the confidence interval with coefficient 1-α from the test with significance level α by defining the interval to be the set of all values of θ_0 that would be accepted by the test on the basis of the data. We shall see this correspondence repeatedly illustrated in the next two chapters.

5. Post-test estimation

As a general principle, a test of hypothesis should be followed by a decision as to what value to use for the unknown parameter in any future work. When the null hypothesis is simple (that is, fully specifies the distribution) and of the form

$$H_0 : \theta = \theta_0 ,$$

the 'estimator' of the unknown parameter θ to be used following acceptance of H_0 is θ_0 (we enclose the word estimator in quotation marks since θ_0 will often be a fixed value, a degenerate case of what we normally think of as an estimator). When the null hypothesis is compound, for example,

$$H_0 : \theta \leq \theta_0 ,$$

and our test leads us to accept H_0, an appropriate estimator of θ is $\min(\theta^*,\theta_0)$; for $H_0 : \theta \geq \theta_0$, it is $\max(\theta^*,\theta_0)$.

On the other hand, when our test leads us to reject H_0, whether simple or compound, the best information we have about θ is the estimator θ^* which appeared in our test criterion and so this is what we use for θ subsequently. If H_0 asserts that certain parameters are equal, thus

$$H_0 : \theta_1 = \theta_2 = \ldots = \theta_k = \theta_0 \quad \text{say,}$$

but θ_0 is unknown, a suitable estimator of θ_0 (and, therefore, of all the θ's) when the test leads to acceptance of H_0 is that obtained by pooling the individual estimators θ^*'s of the θ's. By pooling, we mean that we form a suitably weighted average of the individual θ^*'s. When the test leads to the rejection of H_0, appropriate estimators of the θ's are the individual θ^*'s.

A summary of the estimators to be used in various situations is given below.

(i) Tests on a single mean

H_0	Estimator of μ on acceptance of H_0
$\mu = \mu_0$	μ_0
$\mu \leq \mu_0$	$\min(\mu_0, \bar{X})$
$\mu \geq \mu_0$	$\max(\mu_0, \bar{X})$

In all cases, use $\bar{X}$ as an estimator of μ if H_0 is rejected.

(ii) Tests comparing two means

$$H_0 : \mu_1 - \mu_2 = \mu_0 .$$

Accept H_0 : use estimator μ_0 for $(\mu_1 - \mu_2)$; we suggest

use estimator $(\Sigma X_i + \Sigma Y_i)/n + \mu_0/2$ for μ_1,

use estimator $(\Sigma X_i + \Sigma Y_i)/n - \mu_0/2$ for μ_2.

Reject H_0 : use estimator $(\bar{X} - \bar{Y})$ for $(\mu_1 - \mu_2)$,

use estimator $\bar{X}$ for μ_1,

use estimator $\bar{Y}$ for μ_2.

Here $\bar{X} = \Sigma X_i/n_1$, $\bar{Y} = \Sigma Y_i/n_2$, $n = n_1 + n_2$, where n_1 and n_2 are the respective sample sizes.

The estimators for μ_1 and μ_2 when H_0 is accepted illustrate the point made earlier about pooling individual estimators. When μ_0 is zero so that H_0 asserts that the two distribution means are equal, the best estimator for μ_1 and μ_2 is $(\Sigma X_i + \Sigma Y_i)/n = (n_1\bar{X} + n_2\bar{Y})/(n_1 + n_2)$, the grand mean of all the observations which is the average of the two sample means $\bar{X}$ and $\bar{Y}$ weighted by the number of observations in each sample.

(iii) Tests comparing several means

$$H_0 : \mu_1 = \mu_2 = \ldots = \mu_k = \mu \text{ , say .}$$

Accept H_0 : use estimator $S/n = \sum_{ij}\sum X_{ij}/\Sigma n_i$, the grand mean, for μ.

Reject H_0 : use estimator $S_i/n_i = \bar{X}_i$ for μ_i $(i = 1,2,\ldots k)$.

(iv) Tests on a single variance

H_0	Estimator of σ^2 on acceptance of H_0
$\sigma^2 = \sigma_0^2$	σ_0^2
$\sigma^2 \leq \sigma_0^2$	$\min(\sigma_0^2, s^2)$
$\sigma^2 \geq \sigma_0^2$	$\max(\sigma_0^2, s^2)$

In all cases, use s^2 as an estimator of σ^2 if H_0 is rejected.

(v) Comparing two variances

$$H_0 : \sigma_1^2 = \sigma_2^2 .$$

Put $s^2 = \left\{\Sigma(X_i - \bar{X})^2 + \Sigma(Y_i - \bar{Y})^2\right\}/(n_1 + n_2 - 2)$,

$$s_1^2 = \Sigma(X_i - \bar{X})^2/(n_1-1) ,$$

$$s_2^2 = \Sigma(Y_i - \bar{Y})^2/(n_2-1) .$$

Accept H_0 : use s^2 as an estimator of both σ_1^2 and σ_2^2 .

Reject H_0 : estimate σ_1^2 by s_1^2 and σ_2^2 by s_2^2 .

(vi) Tests comparing several variances

$$H_O : \sigma_1^2 = \sigma_2^2 = \ldots = \sigma_k^2 .$$

Put $s^2 = \sum_i\sum_j (X_{ij} - \bar{X}_i)^2 / \sum_i (n_i - 1) = \sum_i\sum_j (X_{ij} - \bar{X}_i)^2/(n - k)$,

$$s_i^2 = \sum_j (X_{ij} - \bar{X}_i)^2/(n_i - 1) .$$

Accept H_O : use s^2 as an estimator for all the σ^2's .

Reject H_O : estimate σ_i^2 by s_i^2 $(i = 1,2,\ldots k)$.

CHAPTER 8

TESTS ON MEANS

We shall restrict ourselves here mainly to tests relating to the normal distribution, although in the case of large samples, the data can have any distribution with finite mean and variance. We shall also briefly consider two non-parametric tests which are useful when the distribution of the data is unknown.

1. Tests on the mean of a single sample

We consider three cases corresponding to those of Chapter 6, Section 1 where we discussed confidence intervals. As in that section, we suppose that we have a sample of independent observations $X_1, X_2, \ldots X_n$.

Case 1. Normal distribution with known variance

We assume that our observations are distributed as $N(\mu, \sigma^2)$, where μ is unknown, but σ^2 is known. Suppose we believe that the correct value of μ is μ_0, and we wish to test whether or not our observations are consistent with this.

Suppose we put

$$H_0 : \mu = \mu_0 ,$$

$$H_1 : \mu > \mu_0 .$$

Our estimator of μ is $\bar{X}$ and so we shall reject H_0 if $\bar{X}$ is too large compared with μ_0. We put

$$T = \frac{\bar{X} - \mu_0}{\sigma/n^{\frac{1}{2}}} . \qquad (1)$$

If H_0 applies, $T \sim N(0,1)$ and thus the probability that $T \geq z(1-\alpha)$ is α, $z(1-\alpha)$, as usual, representing the $1-\alpha$ quantile of the distribution of the standard normal variate. If now we identify α with some chosen significance level, possibly 5, 1 or 0.1 per cent, we decide to reject or accept

H_0 according to whether or not the realization of T, viz.

$$t = \frac{\bar{x} - \mu_0}{\sigma/n^{\frac{1}{2}}} ,$$

does or does not exceed the critical value, $z(1-\alpha)$; in short, we reject H_0 if $t \geq z(1-\alpha)$ and accept it if $t < z(1-\alpha)$. Alternatively, we may work out $P(T \geq t|H_0) = 1 - \Phi(t)$, which is our significance probability for rejecting H_0.

In the same way, if our two hypotheses are

$$H_0 : \mu = \mu_0 ,$$

$$H_1 : \mu < \mu_0 ,$$

we reject H_0 if $t \leq z(\alpha) = -z(1-\alpha)$ since the distribution of T is symmetrical about zero, if H_0 is true.

If the alternative hypothesis H_1 merely states that $\mu \neq \mu_0$, then we reject H_0 if $|t| \geq z(1-\alpha/2)$. This test corresponds to the relevant confidence interval given in Chapter 6 (similarly for other two-sided tests discussed in this section and the next).

Example. We return to our sample of nine random numbers drawn from a population of unit variance as described in the corresponding section of Chapter 6. We wish to test the hypothesis

$$H_0 : \mu = 0 ,$$

against the alternative hypothesis

$$H_1 : \mu \neq 0 .$$

Since $\bar{x} = 0.205$, $\sigma = 1$, $n = 9$, we have $t = 0.615$. The value of $z(0.975)$ corresponding to a significance level of 5 percent is 1.96 which is considerably in excess of the observed value of T. Hence, we accept, at the 5 percent significance level, the null hypothesis that the distribution mean of the random number population is zero.

Case 2. Any distribution with finite variance when n is large

Again we take

$$H_0 : \mu = \mu_0 ,$$

$$H_1 : \mu > \mu_0 .$$

With large n, it follows from our discussion in Chapter 6 that the random variable

$$T = \frac{\bar{X} - \mu_0}{s/n^{\frac{1}{2}}} \tag{2}$$

is approximately distributed as N(0,1) if H_0 is true. Hence we reject H_0 if $t \geq z(1-\alpha)$, since $P\{T \geq z(1-\alpha)|H_0\} = \alpha$. As before $P(T \geq t|H_0) = 1 - \Phi(t)$ is our significance probability.

Similarly the other one-sided test and the two-sided test are performed as described in Case 1 with s replacing σ.

Example. The 65 values of ΔH for the reaction between NaOH and HCl provide a useful example. From the data in Table 3.1, we find

$$\overline{\Delta H} = -57.34 \text{ kJ mol}^{-1} ,$$

$$s = 1.65 \text{ kJ mol}^{-1} .$$

From the literature, we find that the value for ΔH is -56.40 kJ mol^{-1}, and it is naturally of interest to compare our value with this to see if the two agree. Now, consideration of the experimental method does not lead us in the first instance to suspect that our results are likely to be biased in any one direction. For this reason, it is therefore appropriate to test the null hypothesis

$$H_0 : \mu = -56.40 \text{ kJ mol}^{-1} ,$$

against the alternative,

$$H_1 : \mu \neq -56.40 \text{ kJ mol}^{-1} .$$

We thus require a two-sided test. The modulus of our test statistic is

$$\left| \frac{-57.34 - (-56.40)}{1.65/65^{\frac{1}{2}}} \right| = |-4.59| = 4.59 \; .$$

Under H_0, the corresponding critical value is 1.96 at the 5 percent significance level. Our value exceeds this so we reject H_0, accepting H_1, that the population mean of our data does not agree with the value quoted in the literature.

In this event, we regard the value of -57.34 kJ mol^{-1} as the best estimate of the enthalpy of neutralization that can be obtained from this particular series of observations. We may be tempted to infer that there is a systematic difference between our experimental method and that employed in the determination of the literature value, causing our mean value to be lower than the literature value. However, such a systematic difference due to method would be surprising - but what can the mean difference be due to, if not to systematic error? We note that the significance probability is $P(|T| \geq 4.59 | H_0) < 10^{-5}$, so clearly we are not at serious risk of rejecting H_0 incorrectly. If we had performed a one-sided test with $H_1 : \mu < -56.40$ kJ mol^{-1}, the result would have been even more significant. One possibility is that there is an appreciable experimental error associated with the literature value, of which we have taken no cognizance. Knowledge of this would enable us to use the test described later in Section 2, Case 4. Unfortunately, the literature value for $-\Delta H$ is quoted without a statement of the experimental error (like so many literature values!). Nevertheless, in this particular case, we may feel reasonably confident that there is very little error associated with the literature value, because of the importance of the enthalpy of neutralization in studies of the physical chemistry of aqueous solutions.

To complete the story, it only remains to trace the source of the difference. Consultation of more extensive tables (Landolt-Börnstein) reveals that ΔH is a function of temperature and the concentrations of HCl and NaOH. The value of -56.40 kJ mol^{-1} refers to the value in the limit of zero concentration at 25°C, whereas our work was carried out at concentrations close to 0.1 M at approx. 20°C. From the tables, the value quoted for conditions nearest to those employed by us is -57.28 kJ mol^{-1}. Clearly then it is not surprising that we have found a value close to this. To check that everything is in order, we now test the hypotheses

$$H_0 : \mu = -57.28 \text{ kJ mol}^{-1} ,$$

$$H_1 : \mu \neq -57.28 \text{ kJ mol}^{-1} ,$$

arguing that there is still no reason to use a one-sided test. The realization of our test statistic comes out at -0.29 with modulus less than the critical value of 1.96 for the 5 percent significance level. Hence we may accept H_0 and conclude that our experimental procedure produces a result in agreement with that in the literature, provided that due account is taken of the effects of temperature and concentration.

Case 3. Normal distribution with unknown variance

In the first instance, we take

$$H_0 : \mu = \mu_0 ,$$

$$H_1 : \mu > \mu_0 .$$

We again compare $\bar{X}$ with μ in some convenient way. We use the test criterion

$$T = \frac{\bar{X} - \mu_0}{s/n^{\frac{1}{2}}} ,$$

which under H_0 is distributed as t_{n-1}. We proceed analogously to the two preceding cases, choosing a significance level α and a critical value of Student's t, say $t_{n-1}(1-\alpha)$, with which to compare the experimental value of T. We reject H_0 if

$$\frac{\bar{x} - \mu_0}{s/n^{\frac{1}{2}}} \geq t_{n-1}(1-\alpha) ,$$

or alternatively we calculate the significance probability

$$P\left\{T \geq \frac{\bar{x} - \mu_0}{s/n^{\frac{1}{2}}} \;\middle|\; H_0\right\}$$

from tabulated values of the distribution function of

Student's t.

For the other one-sided test, we proceed similarly, rejecting H_0 if

$$\frac{\bar{x} - \mu_0}{s/n^{\frac{1}{2}}} \leq t_{n-1}(\alpha) = -t_{n-1}(1-\alpha) .$$

For the two-sided situation,

$$H_0 : \mu = \mu_0 ,$$

$$H_1 : \mu \neq \mu_0 ,$$

we reject H_0 if

$$\left| \frac{\bar{x} - \mu_0}{s/n^{\frac{1}{2}}} \right| \geq t_{n-1}(1-\alpha/2) .$$

NB. Tables to assist in the choice of samples sizes required to have a high probability of a given mean difference being significant may be found, for instance, in Owen's Handbook of Statistical Tables, Section 2.

Example. In benzene solution, n-butyl lithium exists in the form of associated molecules $(C_4H_9Li)_m$, where m is called the degree of association. Six solutions of this compound were studied by the method of vapour pressure lowering, and the following values for m were obtained: 5.980, 6.089, 6.198, 6.151, 6.479, 6.105 . Consideration of the experimental method suggests that the results would be biased on the larger side of the true value and so a one-sided test is appropriate. A simple theory is that the molecules are hexameric and that each observation is made up of the integer 6 together with a small random error and possibly a small systematic error. Thus we can formulate two models for our observations, X.

Model 1, random error only

$$X_i = 6 + \varepsilon_i , \quad \varepsilon_i \sim N(0,\sigma^2), \quad i = 1,2,\ldots 6 .$$

Model 2, random and systematic error

$$X_i = 6 + \eta + \varepsilon_i \ , \ \eta > 0, \ \varepsilon_i \sim N(0,\sigma^2), \ i = 1,2,\ldots 6 \ .$$

To test which of these two models applies, we formulate the two hypotheses

$$H_0 : \mu = 6 \quad \text{(model 1)} \ ,$$

$$H_1 : \mu > 6 \quad \text{(model 2)} \ .$$

We calculate

$$\bar{x} = 6.167 \ ,$$

$$s^2 = 0.0287 \ .$$

Our test statistic is thus

$$\frac{6.167 - 6}{0.169/6^{\frac{1}{2}}} = 2.42 \ .$$

The critical value, $t_5(0.95)$, for a one-sided test at the five percent significance level is 2.02. Hence we reject H_0 in favour of the alternative hypothesis, that our method contains a source of systematic error.

2. Tests between means drawn from two samples

We again follow the pattern established in Chapter 6. We suppose that we have two samples $X_1, X_2, \ldots X_{n_1}$ and $Y_1, Y_2, \ldots Y_{n_2}$, all observations being independent. We write μ_1 and μ_2 for the distribution means of the X's and the Y's, and where appropriate σ_1^2 and σ_2^2 for the corresponding variances. We have four cases exactly equivalent to the four cases of Chapter 6, Section 2, and one additional case of matching pairs where X_1 goes with Y_1 , X_2 with Y_2 , and so on.

Case 1. Normal distributions with known variances

We commence by examining the hypothesis

$$H_0 : \mu_1 - \mu_2 = \mu_0 \text{ (usually 0)},$$

against the alternative hypothesis

$$H_1 : \mu_1 - \mu_2 > \mu_0 .$$

Under H_0, the statistic

$$T = \frac{(\bar{X}-\bar{Y}) - \mu_0}{(\sigma_1^2/n_1 + \sigma_2^2/n_2)^{\frac{1}{2}}} \qquad (1)$$

is distributed as N(0,1). Hence, we reject H_0 at the α level of significance if $t \geq z(1-\alpha)$.

If the alternative hypothesis is

$$H_1 : \mu_1 - \mu_2 < \mu_0 ,$$

we reject H_0 if $t \leq - z(1-\alpha)$. If the alternative hypothesis is

$$H_1 : \mu_1 - \mu_2 \neq \mu_0 ,$$

we reject H_0 if $|t| \geq z(1-\alpha/2)$.

In each of the above three situations, we may prefer to quote a significance probability.

We shall not illustrate this test since, as we have explained already, this situation is unusual.

Case 2. Any distribution with finite variances when n_1 and n_2 are large

Again this is the same as Case 1, simply replacing σ_1^2 by s_1^2 and σ_2^2 by s_2^2 in equation (1), that is replacing the unknown distribution variances by the corresponding sample variances.

Example. A good example is provided by the data of the 1967 and 1968 classes for the enthalpy of neutralization of NaOH and HCl. (Tables 3.1 and 6.2). An obvious question is whether or not there is any systematic difference between the results produced by these two classes. Formally, we have to test the hypothesis

$$H_0 : \mu_{1968} - \mu_{1967} = 0 \text{ (no difference)} ,$$

against the alternative

$$H_1 : \mu_{1968} - \mu_{1967} \neq 0 \text{ (there is a difference)}.$$

Note that the alternative hypothesis H_1 simply asserts that the two distribution means are different but makes no statement as to the sign of this difference. This is not an omission but a statement of our belief that there is no reason why one class should produce results above or below those of the other class.

We collect together the relevant data:

(1968)	(1967)
$\bar{x} = 57.34$	$\bar{y} = 56.99$
$s^2(x) = 2.726$	$s^2(y) = 1.130$
$n_1 = 65$	$n_2 = 32$

where the x's and y's stand for the unsigned dimensionless quantities in the tables. Our test statistic is

$$t = \frac{57.34 - 56.99}{(2.726/65 + 1.130/32)^{\frac{1}{2}}}$$

$$= 1.26 .$$

The critical value $z(0.975)$ corresponding to a significance level of 5 percent is 1.96. Hence $t < z(0.975)$ and we accept H_0, thus concluding that there is no significant difference between the results of the two classes. In accordance with Chapter 7, Section 5, we now estimate μ , the mean common to both populations by the grand sample mean $(57.34 \times 65 + 56.99 \times 32)/97 = 57.22$. We observe that this is very close to the literature value of 57.28 (see previous section).

Case 3. Normal distributions with the same, but unknown, variance

The two hypotheses for test are just as before. We shall deal with the one-sided situation defined by

$$H_0 : \mu_1 - \mu_2 = \mu_0 ,$$

$$H_1 : \mu_1 - \mu_2 > \mu_0 .$$

Our test statistic is

$$T = \frac{(\bar{X}-\bar{Y}) - \mu_0}{s(1/n_1 + 1/n_2)^{\frac{1}{2}}} , \qquad (2)$$

where s^2 is the pooled estimator of the common variance of X and Y, namely

$$s^2 = \frac{\Sigma(X_i-\bar{X})^2 + \Sigma(Y_i-\bar{Y})^2}{n_1 + n_2 - 2} . \qquad (3)$$

Under H_0, T is distributed as Student's t with n_1+n_2-2 degrees of freedom. Hence we reject H_0 at the α level of significance if the realization of T is greater than or equal to $t_{n_1+n_2-2}(1-\alpha)$.

The other one-sided case and the two-sided case are dealt with as before.

NB. Tables to assist in the choice of sample sizes required to have a high probability of a given mean difference being significant may be found, for instance, in Owen's Handbook of Statistical Tables, Section 2.

Example. We use the density data for the two high polymers H1 and H2 which we have tabulated in Tables 5.4 and 6.3. Loosely speaking, we want to know whether or not there is a significant difference between the densities of these two samples. More precisely, we examine our data in relation to the two hypotheses

$$H_0 : \mu_{H1} - \mu_{H2} = 0 ,$$

$$H_1 : \mu_{H1} - \mu_{H2} \neq 0 ,$$

since, as far as we are concerned, there is nothing to choose between our two samples (both being prepared under similar conditions). A two-sided test is clearly appropriate here. First, we assemble together the relevant data using the results of Chapter 6, Section 2, Case 3. We have

(H1)	(H2)
$\bar{x} = 1.21510$	$\bar{y} = 1.21650$
$n_1 = 5$	$n_2 = 5$

$$s = 6.10 \times 10^{-4} .$$

Hence the realization of our test statistic is

$$\frac{(1.21510 - 1.21650)}{6.10 \times 10^{-4} \times (2/5)^{\frac{1}{2}}} = -3.63 .$$

To decide whether or not to reject H_0 we have to compare the modulus of this result, viz. 3.63, with the value of $t_8(1-\alpha/2)$, where α is our chosen significance level. If we take α as 0.05, we use $t_8(0.975) = 2.31$. Hence, we reject H_0 and base future action on the supposition that there is a real difference between the densities of the two polymers.

Since we had a very strong prior belief that H_0 was true, we might wish to lower our significance level. However, our t value is even significant at the one percent level, being greater than $t_8(0.995) = 3.36$. The evidence against H_0 is thus quite strong, and it seems appropriate to reject H_0 and seek to discover the reason for the difference in density of the two polymers. Following Chapter 7, Section 2, we now estimate μ_{H1} by $\bar{x} = 1.21510$, and μ_{H2} by $\bar{y} = 1.21650$.

Case 4. Normal distributions with unequal unknown variances

We consider the one-sided situation corresponding to the two hypotheses

$$H_0 : \mu_1 - \mu_2 = \mu_0 ,$$

$$H_1 : \mu_1 - \mu_2 > \mu_0 .$$

We use the random variable

$$T = \frac{(\bar{X}-\bar{Y}) - \mu_0}{(s_1^2/n_1 + s_2^2/n_2)^{\frac{1}{2}}} , \tag{4}$$

where $s_1^2 = \Sigma(X_i-\bar{X})^2/(n_1-1)$ and $s_2^2 = \Sigma(Y_i-\bar{Y})^2/(n_2-1)$. Under H_0, T is approximately distributed as Student's t with ν degrees of freedom, where ν, as explained in the corresponding section of Chapter 6, is a weighted average of $\nu_1 = n_1-1$ and $\nu_2 = n_2-1$; thus

$$\frac{1}{\nu} = \frac{1}{\nu_1}\left(\frac{s_1^2/n_1}{s_1^2/n_1 + s_2^2/n_2}\right)^2 + \frac{1}{\nu_2}\left(\frac{s_2^2/n_2}{s_1^2/n_1 + s_2^2/n_2}\right)^2 . \tag{5}$$

We reject H_0 at the significance level α if the realization of T is greater than or equal to $t_\nu(1-\alpha)$.

The other one-sided case and the two-sided case are dealt with as before.

Example. We complete the parallel of this section with that of Section 2 in Chapter 6 by re-examining the data obtained by two different methods for the percentage of a constituent W in solution.

The relevant numerical quantities are

(method A)	(method B)
$\bar{x} = 5.140$	$\bar{y} = 5.203$
$s_1^2 = 11.11 \times 10^{-4}$	$s_2^2 = 45.34 \times 10^{-4}$
$n_1 = 10$	$n_2 = 10$
$\nu_1 = 9$	$\nu_2 = 9$

$$\nu = 13.2 .$$

The two alternative hypotheses are

$$H_0 : \mu_A - \mu_B = 0 ,$$

$$H_1 : \mu_A - \mu_B \neq 0 .$$

The realization of our statistic T is -2.65 and its modulus is therefore 2.65. For a significance level of 5 percent, we have $t_{13.2}(0.975) = 2.16$. Hence we conclude that there is a significant difference between the two methods and reject H_0. As a result we cannot say what is the percentage of W in the solution, since the two methods yield different means, but we can estimate this mean difference by $\bar{y}-\bar{x} = 0.063$.

Case 5. Matching pairs from two normal distributions with unknown variances

It is sometimes helpful to design an experiment so that the effect of one possible source of variation is cancelled out in any comparison to be made. For example, in comparing the effect of a new diet as compared with a standard diet on the weight gain of rats during a specified period of time, we may remove some of the variation between rats by randomly selecting one rat from each of several litters to have the standard diet and another rat from the same litter to have the new diet. In this way the variation between weight gains due to litter differences is cancelled out when comparing the diets, assuming the litter differences have an additive effect.

Let $X_1, X_2, \ldots X_n$ be one set of values (the weight gains under the standard diet in the above example) and $Y_1, Y_2, \ldots Y_n$ the other set of values (weight gains under the new diet), arranged so that $\{X_i, Y_i;\ i = 1 \text{ to } n\}$ form a set of matching pairs. We take the differences $D_i = X_i - Y_i$ for all i, and use these as a single sample of size n to perform the test of Case 3 under Section 1. For a one-sided test, the two hypotheses are

$$H_0 : E(D) = \mu_1 - \mu_2 = \mu_0 ,$$

$$H_1 : E(D) = \mu_1 - \mu_2 > \mu_0 .$$

The other one-sided test and the two-sided test are formulated similarly.

This test loses some power as compared with the ordinary Case 3 two-sample test discussed in Section 2 owing to the test criterion having only n-1 degrees of freedom here, but twice that number when treated as a two-sample situation. However, this effect would ordinarily be overwhelmed by the influence of the other sources of variation (in our example, the variation in the weight gains arising from litter differences), which would thus make the two-sample procedure much less powerful than the present one.

Example. As an illustration, we take some data on the effect of an intramuscular injection of the steroid Betamethasone on the glucose entry rate expressed as mg carbon min^{-1} into the blood plasma of sheep. These data are given in Table 8.1. [Reilly and Ford (1974)]

Table 8.1. Rate of glucose entry into the blood plasma of sheep before and 24 hour after injection with Betamethasone

Sheep No.	x = Rate of glucose entry/(mg min^{-1})	
	Before	After
1	23.0	40.0
2	30.0	34.5
3	16.0	20.1
4	17.0	20.0
5	36.0	56.0
6	29.0	38.0

Previous work leads us to expect μ_{after} to be larger than μ_{before}, so that we propose the following hypotheses:

$$H_0 : \mu_{before} - \mu_{after} = 0 ,$$

$$H_1 : \mu_{before} - \mu_{after} < 0 .$$

We define

$$D = X_{before} - X_{after} .$$

We calculate

$$\bar{d} = -9.6 ,$$

$$s = 7.3 .$$

Our test criterion is thus

$$\frac{-9.6 - 0}{7.3/6^{\frac{1}{2}}} = -3.2 .$$

Now the value of $t_5(0.95)$ appropriate to a one-sided test at the five per-cent level is 2.02. We see that the value of our test statistic is less than -2.02 and so we conclude that the rate of glucose entry into the blood plasma is significantly enhanced after injection of the steroid, and reject H_0. Thus we estimate $\mu_{after} - \mu_{before}$ as 9.6.

3. Tests on proportions

Here we are thinking of large sample sizes so that we may use the normal approximation to the binomial distribution, although for small samples we may use the binomial distribution itself.

3.1. Single sample

Suppose we have $K \sim B(p,n)$ and we wish to test between

$$H_0 : p = p_0 ,$$

$$H_1 : p > p_0 .$$

Writing $\hat{p} = K/n$, we use the test criterion

$$T = \frac{\hat{p} - p_0}{\{p_0(1-p_0)/n\}^{\frac{1}{2}}} , \qquad (1)$$

which is approximately distributed as N(0,1) under H_0. We thus reject H_0 at the α level of significance if our observed t exceeds $z(1-\alpha)$. If n is not very large, it is a little more accurate to use the half correction with test criterion

$$T = \frac{K-\frac{1}{2}-np_0}{\{np_0(1-p_0)\}^{\frac{1}{2}}} . \qquad (2)$$

The tests for the other one-sided situation and the two-sided situation are analogous. The two-sided test is equivalent to the confidence interval given by the roots of equation (3.1.3) of Chapter 6, whereas if we replaced p_0 in the denominator of equation (1) by $\hat{p}$ we would obtain the test equivalent to the interval derived from equation (3.1.1) of Chapter 6.

The sign test in Section 4.1 is a particular case of this test with $p_0 = \frac{1}{2}$.

Example. We return to the example of Section 3.1 in Chapter 6. We wish to test the null hypothesis of complete ignorance using a one-sided test:

$$H_0 : p = 0.25 ,$$

$$H_1 : p > 0.25 .$$

The realization of our test criterion is

$$t = \frac{0.30 - 0.25}{(0.25 \times 0.75/40)^{\frac{1}{2}}}$$

$$= 0.730 .$$

For a one-sided test at the 5 percent level the critical value is $z(0.95) = 1.64$ and hence we accept H_0.

3.2. Two samples

Suppose $K_1 \sim B(p_1,n_1)$, $K_2 \sim B(p_2,n_2)$ and K_1, K_2 are independent, and that we wish to test between the two hypotheses:

$$H_0 : p_1 - p_2 = p_0 ,$$

$$H_1 : p_1 - p_2 > p_0 .$$

Writing $\hat{p}_1 = K_1/n_1$, $\hat{p}_2 = K_2/n_2$, $\hat{q}_1 = 1-\hat{p}_1$, $\hat{q}_2 = 1-\hat{p}_2$, we use the test criterion

$$T = \frac{(\hat{p}_1-\hat{p}_2) - p_0}{(\hat{p}_1\hat{q}_1/n_1 + \hat{p}_2\hat{q}_2/n_2)^{\frac{1}{2}}} ,$$

which is approximately distributed as N(0,1) under H_0 . We reject H_0 at the significance level α if $t \geq z(1-\alpha)$. Again, the other one-sided and the two-sided test follow in the usual way. Again, also, the two-sided test is equivalent to the confidence interval given in Chapter 6, Section 3.2.

Example. Using the data in the example of Section 3.2 of Chapter 6, we test the null hypothesis that there is no difference between the proportion of voters of Scotland and Northern Ireland voting 'yes'. Supposing no prior information about the preferences of the voters of the two countries, we use a two-sided test:

$$H_0 : p_1 - p_2 = 0 ,$$

$$H_1 : p_1 - p_2 \neq 0 .$$

The realization of our test statistic is

$$t = \frac{0.58430 - 0.52146}{\{(1.065 + 5.019) \times 10^{-7}\}^{\frac{1}{2}}}$$

$$= 80.56 ,$$

which exceeds z(0.9995) = 3.29. Hence we reject H_0 at the 0.1 percent level and conclude that there is a significant difference between the preferences of the voters in Scotland and Northern Ireland.

4. Nonparametric tests

There are useful tests of location (e.g. mean) when the distribution of the data is unknown. We shall consider two nonparametric tests in this section, the Sign test, and the Wilcoxon (or Mann-Whitney) test. The Sign test is quick and easy to use for a 'matching pairs' situation (see Case 5 of

Section 2) with small sample size n, or reasonably so for large n; at intermediate values of n (say, $10 < n < 20$), it is less convenient. It may also be used for a one-sample test. However, it is not very powerful, while the Wilcoxon test generally has good power.

4.1. The Sign test

For the Sign test, we compare n matched pairs of X and Y values, and note in how many cases X exceeds Y (or, how many of the differences $X_i - Y_i$ are positive - hence the name of the test). Under the null hypothesis, this number, K say, is distributed as B(0.5,n). We test by seeing if the realization k is too far away from n/2 in a direction indicated by H_1 for us to accept H_0 at our significance level. For small n , we cannot usually choose just any significance level in advance, as will become clear.

Suppose $K \sim B(p,n)$ and we have

$$H_0 : p = 0.5 ,$$

$$H_1 : p > 0.5 .$$

If we were to make a rule to reject H_0 for $k \geq c$, where c is some integer, this would correspond to a significance level of

$$P(K \geq c) = 2^{-n}\left\{\binom{n}{n} + \binom{n}{n-1} + \binom{n}{n-2} + .. \binom{n}{c}\right\} \tag{1}$$

where $\binom{n}{r} = {}^nC_r$. Hence

$$P(K \geq c) = 2^{-n}[1 + n + n(n-1)/2 + \ldots \{n(n-1)\ldots(n+1-c)\}/c!].$$

With the same H_0 , but with $H_1 : p < 0.5$, it is appropriate to reject H_0 for $k \leq c$ for some c with significance level

$$P(K \leq c) = 2^{-n}\left\{\binom{n}{0} + \binom{n}{1} + \binom{n}{2} + .. \binom{n}{c}\right\} \tag{2}$$

$$= 2^{-n}[1 + n + n(n-1)/2 + \ldots \{n(n-1)\ldots(n+1-c)\}/c!].$$

We notice that the corresponding probability terms are the same in the two tails; this is because the distribution $B(0.5,n)$ is symmetrical about $n/2$. For the same H_0 again, and $H_1 : p \neq 0.5$, we reject for $|k-n/2| \geq c$, for some c, where $n/2 - c = c_0$, say, is an integer. Then the significance level of the test is

$$P(|K-n/2| \leq c) = 2P(K \leq c_0)$$

$$= 2^{-n+1}[1 + n + \ldots + \{n(n-1)\ldots(n+1-c_0)\}/c_0!].$$

For each of these tests, c (or c_0) is chosen to give a significance level about the desired value. Alternatively, we may quote the significance probability corresponding to our observed k.

For large n, it is convenient to use the normal approximation, preferably with the half correction (see Chapter 4, Section 6), whereby

$$P(K \geq c) \sim 1 - \Phi\left(\frac{c-\frac{1}{2}-n/2}{n^{\frac{1}{2}}/2}\right), \tag{3}$$

$$P(K \leq c) \simeq \Phi\left(\frac{c+\frac{1}{2}-n/2}{n^{\frac{1}{2}}/2}\right). \tag{4}$$

Incidentally, the Sign test may also be used to test the null hypothesis that δ, say, is the median $\xi_{0.5}$, of the distribution from which we have n observations, using as K the number of observations exceeding δ. (Note: for a symmetric distribution, the mean equals the median).

Example 1. For each of 10 sets of twins a sleeping tablet of type A was given to a randomly selected twin, and a sleeping tablet of type B was given to the other, and the lengths of their ensuing sleeping times were recorded. In 8 of the 10 cases, tablet A produced a longer sleeping time than tablet B. We wish to examine whether this is significant evidence against the null hypothesis that the two tablets are equally effective.

Here we are not supposing a prior expectation that A may give more sleep than B, so a two-sided test is appropriate. The significance probability is

$$2P(K \geq 8 | p = 0.5) = 2 \times 2^{-10}(1 + 10 + 10 \times 9/2)$$

$$= \frac{2}{1024}(11 + 45) = \frac{56}{512} = 0.109 .$$

This probability is rather too high for us to reject H_0. Hence we proceed on the assumption that A and B give the same sleeping times. (Note: if a count of 9 cases favouring A had been recorded, the significance probability would have been 11/512 = 0.021, in which case it would have been more reasonable for us to reject H_0).

Example 2. A particular coin is suspected of yielding more heads than tails when tossed. In an experiment 61 heads resulted from 100 tosses. We wish to test whether or not our suspicion is justified. We have

$$H_0 : p = 0.5 ,$$

$$H_1 : p > 0.5 .$$

The significance probability is

$$P(K \geq 61) \simeq 1 - \Phi\left(\frac{61-0.5-50}{5}\right) = 1 - \Phi(2.1) = 1 - 0.9821$$

$$= 0.0179 .$$

So we reject H_0 in favour of H_1 at all significance levels greater than about 1.8 percent.

4.2. The Wilcoxon (Mann-Whitney) test

Suppose $X_1, X_2, \ldots X_{n_1}$ are independently distributed as $F(x)$, and $Y_1, Y_2, \ldots Y_{n_2}$ are independently distributed as $G(y)$. We may use the Wilcoxon test, which is a kind of location test, equivalent to a test between two means, but without assuming particular distributions. In this test, we assume that

$$G(x) = F(x-\theta)$$

and the null and alternative hypotheses are given as

$$H_0 : \theta = 0 ,$$

$$H_1 : \theta > 0 \text{ (or } H_1 : \theta < 0, \text{ or } H_1 : \theta \neq 0) .$$

We use the test criterion S, equal to the sum of ranks of the X's in the combined, ordered sample of X's and Y's. We may regard either sample as being the X's, and it is natural to choose that sample which will involve fewer or easier additions (depending on n_1, n_2, and the relative positions of the X's and Y's). It can easily be shown that an equivalent test criterion is that proposed by Mann and Whitney, namely U, equal to the sum for each X of the number of Y's it exceeds, summed over the X's. In fact $U = S - n_1(n_1+1)/2$. It is easier to use U for calculation purposes, whether working by hand or using a computer. It can also easily be shown that under H_0, the distribution mean of U is $n_1n_2/2$ and the variance is $n_1n_2(n+1)/12$, where $n = n_1+n_2$. Clearly under H_0, S has the same variance as U, but mean $n_1(n+1)/2$. Tables of the distribution of U (and S) are found in Owen's "Handbook of Statistical Tables", and many other places. However, the distributions of U and S are approximately normal for large n_1, n_2; indeed this approximation is satisfactory for n_1, $n_2 > 8$, which will often be true.

If H_1 asserts $\theta > 0$, that is Y tends to be larger than X, then using the normal approximation, we reject H_0 for a significantly small value of U, that is for $t \leq z(1-\alpha)$, where

$$T = \frac{U + \frac{1}{2} - n_1n_2/2}{\{n_1n_2(n+1)/12\}^{\frac{1}{2}}} \tag{1}$$

or alternatively quote the significance probability for an observed value u as

$$\Phi\left(\frac{u + \frac{1}{2} - n_1n_2/2}{\{n_1n_2(n+1)/12\}^{\frac{1}{2}}}\right) .$$

In both forms of the test, we have also incorporated the half correction for continuity which could be omitted if it is small compared to $U - n_1n_2/2$.

As in previous sections, we perform the other one-sided or the two-sided test in the appropriate way, where the half correction should be included in such a way as to reduce the significance probability, or, equivalently, to reduce $|T|$.

We have been assuming that the data are continuous, in which case the probability of a tie (two or more identical observations) will be zero. However, real data are never strictly continuous since we only record observations to a finite number of decimal places. In practice, therefore, **ties** sometimes occur. The usual procedure to cater for ties is to add ½ for each Y value equal to an X value when calculating U, or to use the average rank for each tied observation when S is being used. If, as a result of this correction for ties, our U or S value is not an integer, we would not use the half correction for the normal approximation. We may also correct the variance when ties are involved by multiplying it by the factor

$$\{1 - \Sigma(t_i^3-t_i)/(n^3-n)\} ,$$

where t_i is the multiplicity of the ith tie, but usually this correction is negligible and may reasonably be omitted.

The Wilcoxon (or Mann-Whitney) test is known to have high power in many circumstances. When F and G are both normal distribution functions with the same variance, an appropriate asymptotic (large n_1, n_2) measure of the power of the test relative to the optimum t test is 95.5 percent, which is very high for conditions favouring the t test. The measure is over 100 percent with certain other distributions, and has been shown to be never less than 86.4 percent.

Example. Here we exemplify the test using U and the normal approximation to the distribution of T. Suppose that the marks in a particular botany examination obtained by 14 girls and 10 boys were as follows (after arranging in order of size).

G:		34		39	43		50				56
B:	31		35	39		45		51	52	54	56
G:			61	65	68	70	71	73	76	79	83
B:	58	60									

We examine whether these data afford significant evidence that, among girls and boys represented by this sample, the girls are better at botany than the boys.

For convenience regarding the B values as X's, we sum the 10 contributions to U. The boy with 31 marks has the lowest mark in the whole sample and thus contributes 0 to U; the boy with 35 marks is superior to only one of the girls and he thus contributes 1 to U; the next boy has one girl below him in marks and one girl tied with him - he thus contributes $1 + \frac{1}{2}$ to U. Continuing in this way, we obtain

$$U = 0 + 1 + 1\tfrac{1}{2} + 3 + 4 + 4 + 4 + 4\tfrac{1}{2} + 5 + 5 = 32 .$$

From the way in which we posed the question under consideration our hypotheses are

$$H_0 : \mu_G = \mu_B ,$$

$$H_1 : \mu_G > \mu_B ,$$

where μ_B and μ_G are the means for boys and girls. Under H_0 , U is approximately distributed as $N\{n_1n_2/2, n_1n_2(n+1)/12\}$, and here $n_1 = 10$, $n_2 = 14$, $n = 24$, so that U is approximately distributed as N(70,291.6). The significance probability is

$$P(U \leq 32) \simeq \Phi\left(\frac{32 + 0.5 - 70}{291.6^{\frac{1}{2}}}\right)$$

$$= \Phi(-2.20)$$

$$= 1 - 0.9861$$

$$= 0.0139 .$$

Hence, we reject H_0 in favour of H_1 at any significance level greater than about 1.4 percent. We conclude that there is significant evidence in favour of the superiority of the girls.

5. Tests between several means

We have already examined ways of testing for a systematic difference between two sets of observations. Here, we consider the more general situation where we have observations from say, k different sources, or k different treatments, or k different levels of a single factor. Consequently, we shall test whether or not the corresponding k means are significantly different. We shall assume throughout this section that the observations are independently, normally distributed with the same variance.

A common procedure is to make the first set of observations, say n_1 altogether, using the first treatment, the second set of observations, say n_2 altogether, using the second treatment, and so on. We thus end up with k sets of observations containing n_1 in the first, n_2 in the second,etc. such that $n_1 + n_2 + \ldots n_k = n$, the total number of observations. We could then compute the mean of the observations within each set and test whether or not there was any systematic difference between the means of these sets. However, here a difficulty arises if we find as a result of the test that there is a systematic difference between the means of the various sets. It does not follow that this difference arises solely as a consequence of the different treatments since there may be some other factor which is unknown and which is changing and influencing the observations; because of the systematic way in which the different treatments were allocated among the total number of observations, the observed difference in means may be partly or even wholly attributable to the influence of this unknown factor.

An example will make our point clearer. Suppose we wished to test whether or not the e.m.f.'s of Weston cells supplied by a series of manufacturers A to E, were the same or different. We could purchase, say, 6 of each type, insert them one at a time into a potentiometer circuit, and measure the length of potentiometer wire whose potential drop was just the same as that across the cell in question - the 'balance points'. If we were to make our measurements in batches, that is, studying manufacturer A's cells first, then B's and so on, we might well observe a significant difference between the means of the balance points and go on to conclude that the e.m.f. of a Weston cell depends on who makes it. If no precautions had been taken to check the constancy of the potential drop across the total length of the potentiometer wire, a very likely cause of the difference would be the drift in this potential. Of course, the experienced experimentalist would not get into this difficulty since he would monitor the potential drop across the whole wire at intervals throughout the series of tests. Nevertheless, even he would benefit by studying the cells from the various manufacturers in random order as in Case 1.

Case 1. Completely randomized observations

In this case, we randomly allocate the k different treatments (or factor levels, or sources) between n units, the ith treatment being applied to n_i of these, where $n_1 + n_2 + \ldots n_k = n$. For this purpose, we may use a table of random numbers according to the procedure described in Appendix 1. In this way, the results for the different treatments are unbiased. This is the situation appropriate to a one-way Analysis of Variance (one-way ANOVA).

In formulating our two hypotheses H_0 and H_1, it is convenient here to use a slightly different notation to that used earlier. For the jth observation of the ith treatment we have

$$H_0 : X_{ij} = \mu + \varepsilon_{ij} \quad ,$$

$$H_1 : X_{ij} = \mu_i + \varepsilon_{ij} \ ,$$

where the ε's are independently distributed as $N(0,\sigma^2)$. We note that, under H_0 , the distribution means are the same for all treatments whereas they are different under H_1.

Such a model is called a fixed-effects model, where we have a fixed and definite set of k categories associated with the k different treatments. Thus, in the case of the example just discussed, the fixed-effects model states quite simply that the e.m.f. of a Weston cell from manufacturers A to E is the same or that it varies within this particular chosen group of makers. In a random-effects model (and we are not discussing this in detail here), the k different categories are assumed to be a random sample drawn from a population of such categories and, under H_1, instead of thinking in terms of individual category means, we think of a variance between the different possible categories. Because we have chosen k particular categories, we are assuming a fixed-effects model and so the conclusions will relate to the particular categories considered only. Someone may consider the categories to be representative of a wider set of possible categories and may seek to apply his conclusions to the wider set, but strictly they relate only to the categories actually involved in the analysis. Thus our investigator studying the e.m.f.'s of Weston cells produced by manufacturers A to E might wish to extend his conclusions to cover all possible manufacturers. If he were to accept H_0 as a result of the procedure described below, he might well draw the erroneous conclusion that all manufacturers produce a Weston cell with the same e.m.f.; however, all that he can assume is that manufacturers A to E produce a cell with the same e.m.f.

The assumption of normality can be relaxed a little inasmuch as the conclusions drawn in the Analysis of Variance are still approximately true for reasonable departures from normality. However, the assumption of constant variance is most important and is critical to the analysis. Because of this, it is sometimes appropriate to transform the data so as to stabilize the variance; appropriate transformations for different circumstances are discussed in Chapter 9, Section 4.

For a one-way analysis of variance, we do not need to have equal numbers of observations in all the categories.

However, for a multi-way analysis (exemplified briefly under Case 2), it is much more convenient to have the same number of observations in each cell, that is in each combination of categories.

The analysis

Basically what we do is to compare two different estimators of the variance of the observations, using an F test. One estimator is based upon the variation between the means of the data corresponding to the k different categories (the numerator of our test criterion, T), the other is based upon the variation within categories about the category means (the denominator). Whether or not the distribution means for the different categories are different, the denominator of T will be an unbiased estimator of the variance, σ^2; the other estimator, however, will only be unbiased if the category distribution means are equal, otherwise its expectation will be larger than σ^2.

We have n_i observations, X_{ij} , for the ith category ($i = 1,2,\ldots k$; $j = 1,2,\ldots n_i$; $\Sigma n_i = n$, say). We put

$$S_{x,i} = \sum_j X_{ij} ,$$

and

$$S_x = \sum_i S_{x,i} = \sum_i\sum_j X_{ij} \text{ , the sum of all the observations.}$$

The variations we wish to compare are proportional to $\sum_i n_i(X_{i.}-X_{..})^2$ between categories and $\sum_i\sum_j(X_{ij}-X_{i.})^2$ within categories, the overall variation being proportional to $\sum_i\sum_j(X_{ij}-X_{..})^2$. Here $X_{i.} = S_{x,i}/n_i$, the mean of the ith category, and $X_{..} = S_x/n$, the grand mean. These quantities are conveniently obtained by calculating the three differences:

$$\sum_i S^2_{x,i}/n_i - S^2_x/ n ,$$

$$\sum_i\sum_j X_{ij}^2 - \sum_i S_{x,i}^2/n_i \ ,$$

$$\sum_i\sum_j X_{ij}^2 - S_x^2/n \ .$$

For these we calculate the three quantities:

$$\sum_i\sum_j X_{ij}^2 \ , \qquad \sum_i S_{x,i}^2/n_i \ , \qquad S_x^2/n \ .$$

Let ss, ms and df represent sum of squares, mean square and degrees of freedom, respectively. Each mean square, U and V, is given by dividing ss by the corresponding df. We thus form Table 8.2.

Table 8.2. One-way ANOVA

Source of variation	df	ss	ms
Between categories	k-1	$\sum_i n_i(X_{i.}-X_{..})^2 = \sum_i S_{x,i}^2/n_i - S_x^2/n$	U
Within categories	n-k	$\sum_i\sum_j (X_{ij}-X_{i.})^2 = \sum_i\sum_j X_{ij}^2 - \sum_i S_{x,i}^2/n_i$	V
Total	n-1	$\sum_i\sum_j (X_{ij}-X_{..})^2 = \sum_i\sum_j X_{ij}^2 - S_x^2/n$	-

We use the test statistic T given by

$$T = U/V \ ,$$

which, under H_0, is distributed as $F_{k-1,n-k}$. We reject H_0 if $t \geq F_{k-1,n-k}(1-\alpha)$, where α represents the chosen significance level. This, of course, is a one-sided test, since under H_0 $E(U) = E(V) = \sigma^2$, while under H_1 $E(U) > E(V) = \sigma^2$.

When k = 2, this test is actually equivalent to the t-test between two means given in Section 2, Case 3, F here being equal to the square of t there, so that this one-sided test corresponds to the two-sided test described there.

Identifying the significant differences

Suppose we have carried out a one-way ANOVA, obtaining a significant result and we wish to know which of the k(k-1)/2 differences between k means are significant, should we simply perform the k(k-1)/2 corresponding t tests? The answer is no! There is a high probability that some of these tests would yield a significant result even if no real differences existed between the distribution means, and the tests are not independent. We will give just two of quite a selection of tests that could be used here. Incidentally their use is not confined to the one-way situation; the means corresponding to the different levels of one factor in a larger analysis can be compared this way.

Tukey's test

The difference between any two from a set of k means $\bar{Y}_i$ and $\bar{Y}_j$ each based on m observations and with basic error estimator s^2 (in the present context, V of Table 8.2, or W of Table 8.4) with ν degrees of freedom is tested using the appropriate critical value of the studentized range $(\bar{Y}_{max}-\bar{Y}_{min})/s(\bar{Y})$. Tables of these are available in Owen's Handbook of Statistical Tables, or more conveniently (with more values of ν) in Biometrika Tables for Statisticians. We compare the criterion $t = |\bar{y}_i-\bar{y}_j|$ with the critical value from the table multiplied by $s/m^{\frac{1}{2}}$. The difference is significant if it exceeds this product. This test is possibly the most powerful available one to use when the means are based on the same sample size.

In practice we look at the largest mean difference first and work downwards to the first nonsignificant difference, as exemplified in the following and in the Case 2 examples.

Scheffé's test.

This is really for testing whether or not any contrast of a set of k means $\{\bar{Y}_i\}$ of samples of sizes $\{n_i\}$ is significantly different from zero. A contrast is a linear combination, $H = \Sigma a_i \bar{Y}_i$, such that $\Sigma a_i = 0$. With error estimator s^2 based on ν degrees of freedom, we have $s^2(H) = s^2 \Sigma a_i^2/n_i$. For the contrast $\bar{Y}_i - \bar{Y}_j$ we have $a_i = 1$, $a_j = -1$.

We compare the realization $|h|$ with the quantity $s(H)\{(k-1)F_{k-1,\nu}(1-\alpha)\}^{\frac{1}{2}}$, and reject the null hypothesis that $E(H) = 0$ at the α level of significance if $|h|$ is greater than this critical value.

Example. We refer once again to the enthalpy data given in Table 3.1. These data comprise 13 groups of 5 determinations and we wish to test whether or not there is any significant difference between the results of the 13 groups. Our hypotheses are

$$H_0 : X_{ij} = \mu + \varepsilon_{ij} ,$$

$$H_1 : X_{ij} = \mu_i + \varepsilon_{ij} ,$$

where $\varepsilon_{ij} \sim N(0,\sigma^2)$, i runs from 1 to 13, and j from 1 to 5. To reduce the amount of computation, we subtract 50 from each datum, forming a working variable Y defined by

$$Y = X - 50 .$$

The three quantities we require for our analysis of variance table are:

$$\sum_i \sum_j y_{ij}^2 = 3680.8 ,$$

$$\sum_i s_{y,i}^2/n_i = \sum_i s_{y,i}^2/5 = 3530.216 ,$$

$$s_y^2/n = 3506.319 .$$

We now draw up our analysis of variance table, as Table 8.3. Our test ratio is thus 0.688 which is considerably less than the critical value of

Table 8.3. One-way ANOVA for the enthalpy data of Table 3.1.

Source of variation	df	ss	ms
Between groups	12	23.897	1.991
Within groups	52	150.584	2.896
Total	64	174.481	-

$F_{12,52}(0.95) = 1.94$. (In fact, we need not have referred to an F table, since t was less than 1, and the F critical values are always greater than 1). Hence we accept H_0 that all the measurements are random samples from the same population.

Referring to Chapter 7, Section 2, we see that we now adopt as our estimate of μ the grand mean, $S_x/n = S_y/n + 50 = 57.34$.

Incidentally, we may test between particular mean differences using Tukey's or Scheffé's test, though here we would not expect any of these to be significant since the overall ANOVA has a far from significant result.

Here the largest mean difference is 58.44 - 56.46 = 1.98, and $s^2 = 2.896$, $\nu = 52$, $m = 5$, $k = 13$. Thus for Tukey's test 1.98 is to be compared with 4.91 (from table) x $(2.896/5)^{\frac{1}{2}} = 3.74$ which is clearly a non-significant result.

For Scheffé's test, $h = 1.98$, $s^2(H) = 2.896 \times (2/5) = 1.1584$. Here the mean difference 1.98 is to be compared with the critical value, $1.1584^{\frac{1}{2}}\{12F_{12,52}(0.95)\}^{\frac{1}{2}} = 5.19$, once more yielding a non-significant result.

Case 2. Randomized blocks of observations

In Case 1 of this section, we supposed that the k treatments were randomly allocated to n units so that part of the observed variation may be due to the different effects of the different units on the observations. For example, if we wish to study the effects of a number of diets on the weight gains of rats, then part of the observed variation could arise as the result of the different parentages.

Where it is possible, we may apply all k treatments to one block (litter, person, plot, etc) randomly ordering the positions of the different treatments within each block in such a way that the variability arising from uncontrolled factors is small within blocks, though possibly large between blocks. Thus, in the case of our rats, we may apply k different diets to different rats from the same litter, for each of several litters, m say. Other examples are (i) applying different suntan lotions to different parts of each of several human backs to compare their degree of protection against sunburn, (ii) applying each of several fertilizers to different subplots of each of several fairly homogeneous plots to compare their effects on crop yield.

Comparing the different treatments within blocks in this way removes extraneous variation which helps to make the test more powerful. This is partially offset by the fact that the number of degrees of freedom associated with the residual (see Table 8.3) is reduced to (m-1)(k-1) from n-k for the completely randomized procedure of Case 1; m, here, stands for the number of blocks and clearly mk = n. However, if the variability of the effects of the different blocks on the observations is large, the net effect will be an appreciable improvement in power. Incidentally the following procedure also applies if we have different levels of a second factor instead of blocks (two-way ANOVA).

The analysis. Here our hypotheses are

$$H_0 : X_{ij} = \mu + \nu_j + \varepsilon_{ij} ,$$

$$H_1 : X_{ij} = \mu + \mu_i + \nu_j + \varepsilon_{ij} ,$$

where μ is the overall effect, μ_i is the ith treatment effect, and ν_j is the jth block effect. We assume, as before, that the ε's are independently distributed as $N(0,\sigma^2)$. We note that, under H_0, an observation X_{ij} is the sum of three parts, the first an overall effect, the second a constant characteristic of a block but independent of the treatment, and the third a random variable. The alternative hypothesis, on the other hand, postulates that the 'block effect' ν_j is simply added to the effect of the treatment $\mu + \mu_i$.

In other respects, we use the same notation as in Case 1 except that we replace $S_{x,i}$ by $S_{x,i.}$ for the sum of the observations for the ith treatment (across the blocks); similarly we let $S_{x,.j}$ represent the total for the jth block. This notation enables us to distinguish the sum of the observations for, say, the third treatment $S_{x,3.}$ from the corresponding sum for the third block $S_{x,.3}$.

We have

$$S_x = \sum_i S_{x,i.} = \sum_j S_{x,.j} = \sum_{ij}\sum X_{ij} \ .$$

We calculate the quantities

$$\sum_{ij}\sum X_{ij}^2 \ , \quad \sum_i S_{x,i.}^2/m \ , \quad \sum_j S_{x,.j}^2/k \ , \text{ and } \quad S_x^2/n \ ,$$

from which we form the sums of squares and mean squares shown in Table 8.4.

We test between the treatments by an F test using as our criterion

$$T = \frac{U}{W} \ .$$

Under H_0, T is distributed as $F_{k-1,h}$, with $h = (m-1)(k-1)$, and so we reject H_0 at the significance level α if $t \geq F_{k-1,h}(1-\alpha)$.

The ratio V/W could be used to test whether the block effects are significantly different, but these are usually of secondary importance, and they are likely to be significantly different anyway.

Table 8.4. ANOVA for randomized blocks

Source of variation	df	ss	ms
Between treatments	k-1	$\sum_i m(X_{i.}-X_{..})^2 = \sum_i S^2_{x,i.}/m - S^2_x/n$	U
Between blocks	m-1	$\sum_j k(X_{.j}-X_{..})^2 = \sum_j S^2_{x,.j}/k - S^2_x/n$	V
Residual	(m-k)(k-1)	$\sum_i\sum_j (X_{ij}-X_{i.}-X_{.j}+X_{..})^2 = \sum_i\sum_j X^2_{ij} - \sum_i S^2_{x,i.}/m - \sum_j S^2_{x,.j}/k + S^2_x/n$	W
Total	n-1	$\sum_i\sum_j (X_{ij}-X_{..})^2 = \sum_i\sum_j X^2_{ij} - S^2_x/n$	-

Example. Four psychological tests, T1, T2, T3 and T4, for assessing I.Q. were compared using five candidates, C1, C2, C3, C4 and C5. The tests were taken by the candidates in random order, with a suitable time interval being allowed between tests. The scores, after subtracting 105 from each, are shown in Table 8.5. We wish to test whether or not the means

Table 8.5. Adjusted scores from I.Q. tests on five candidates

	Candidate						
Test	C1	C2	C3	C4	C5	Total	Mean
T1	-10	-4	1	16	2	5	+1.0
T2	-5	-3	0	20	1	13	+2.6
T3	-1	-2	2	19	5	23	+4.6
T4	-2	-3	0	21	4	20	+4.0
Total	-18	-12	3	76	12	61	

for the different tests are significantly different. We represent the adjusted score by X, and use the subscript i for the tests (the 'treatments') and the subscript j for the candidates (the 'blocks'). Under the null hypothesis, $E(X_{ij}) = \mu + \nu_j$ (the mean score depends on the candidate but not on the test) whereas under H_1, $E(X_{ij}) = \mu + \mu_i + \nu_j$ (the mean score also depends on the test). We calculate the following quantities

$$\sum_i\sum_j X_{ij}^2 = 1677 ,$$

$$\sum_i S_{x,i.}^2/5 = 1123/5 = 224.6 ,$$

$$\sum_j S_{x,.j}^2/4 = 6397/4 = 1599.25 ,$$

$$S_x = 61 ,$$

$$S_x^2/20 = 3721/20 = 186.05 .$$

We thus construct the analysis of variance table shown in Table 8.6.

Table 8.6. Analysis of variance table for I.Q. tests

Source of variation	df	ss	ms
Between tests	3	38.55	12.85
Between candidates	4	1413.20	353.3
Residual	12	39.2	3.26
Total	19	1490.95	-

Our test ratio for a difference between the I.Q. tests is

$$\frac{12.85}{3.26} = 3.93 .$$

Now the value of $F_{3,12}(0.95)$ for a test at the five percent significance level is 3.49 . Hence we reject H_0 and proceed with the supposition that the mean scores for the different tests are different.

Once again we may apply Tukey's and/or Scheffé's tests to the mean differences. This time we would expect to find at least one difference significantly different from zero, since we have found the means for the tests to be significantly different. The six mean differences are as shown (in order of size in each row):

	T3	T4	T2	T1
T3		0.6	2.0	(3.6)
T4			1.4	3.0
T2				1.6

Here $s^2 = 3.26$, $\nu = 12$, $m = 5$, $k = 4$. For Tukey's test we have $s/m^{\frac{1}{2}} = 0.807$, the critical value from the table is 4.20, and the product of these is 3.39, which is therefore the critical value for a mean difference at the 5 percent significance level. Only the ringed, largest difference is significant, that between the means for T3 and T1.

For Scheffé's test, $s(H) = 1.142$, $\{3F_{3,12}(0.95)\}^{\frac{1}{2}} = 3.24$, and the product of these is 3.70, the critical value for Scheffé's test at the 5 percent level here. In this case the largest mean difference is not quite significant. The examples of the use of these tests given in this chapter illustrate the fact that, for means based on equal sample sizes, Tukey's test tends to be more powerful than Scheffé's though the latter can be applied to situations with unequal sample sizes.

CHAPTER 9

TESTS ON VARIANCES

Since a variance affords a standard measure of the spread or dispersion of a probability distribution about its mean, tests on variances are usually used for comparing the spreads of distributions. However, the usual tests on variances assume that the data are normally distributed, and they are not robust to departures from normality as are the tests for means given in the last chapter.

We shall consider here three different types of situation analogous to those of the previous chapter. These situations are the comparison of

1. the variance of a single distribution with a hypothetical value;
2. the variances of two distributions;
3. several variances.

For most of this chapter, we assume that the normal distribution applies, an assumption which is an important requirement for the validity of the tests we shall describe.

1. Test on a single variance

If $X_1, X_2, \ldots X_n$ are our observations, independently, identically distributed as $N(\mu,\sigma^2)$, then the quantity

$$\sum(X_i-\bar{X})^2/\sigma^2 \quad = \quad (n-1)s^2/\sigma^2$$

is distributed as χ^2_{n-1}, as explained in Chapter 4, Section 8. Now if we have some null hypothesis, H_0, which states that the distribution variance is some value σ_0^2, say, it follows immediately that a test criterion T given by

$$T \quad = \quad \Sigma(X_i-\bar{X})^2/\sigma_0^2 \quad = \quad (n-1)s^2/\sigma_0^2$$

is distributed as χ^2_{n-1} under H_0. In the usual way, we have two one-sided and one two-sided test situations according to the way in which we choose to formulate the alternative hypothesis, H_1. The three possibilities are summarised below:

$$H_0 : \sigma^2 = \sigma_0^2 , \begin{cases} \text{(a)} & H_1 : \sigma^2 > \sigma_0^2 , \\ \text{(b)} & H_1 : \sigma^2 < \sigma_0^2 , \text{ or} \\ \text{(c)} & H_1 : \sigma^2 \neq \sigma_0^2 . \end{cases}$$

In case (a), we reject H_0 at the α level of significance if $t \geq \chi^2_{n-1}(1-\alpha)$, where $\chi^2_{n-1}(1-\alpha)$ represents the $1-\alpha$ quantile of the distribution of χ^2_{n-1} . In case (b), we reject H_0 again at the α significance level if $t \leq \chi^2_{n-1}(\alpha)$. In case (c), we reject H_0 if $t \leq \chi^2_{n-1}(\alpha/2)$ or $t \geq \chi^2_{n-1}(1-\alpha/2)$.

Example. We again return to our set of normal random numbers listed in Chapter 6, Section 1. Previously, we have examined whether or not the sample mean of these data was consistent with the supposition that they were drawn from a normal population of unit variance and zero mean, using the theoretical variance of unity in our test. We now test the hypothesis that the true distribution variance of these data is in fact unity. We have no prior indication to adopt a one-sided test and so we seek to test between the hypotheses

$$H_0 : \sigma^2 = 1 ,$$

$$H_1 : \sigma^2 \neq 1 .$$

The sample variance is 1.17 and there are 9 observations in our sample so that the realization of the test criterion is 9.36. Now $\chi^2_8(0.025) = 2.180$ and $\chi^2_8(0.975) = 17.535$. Hence we accept, at the 5 percent significance level, the null hypothesis that the random number population has unit variance.

2. Test on variances from two distributions

Here we assume that we have two sets of observations, $X_1, X_2, \ldots X_{n_1}$, and $Y_1, Y_2, \ldots Y_{n_2}$, each independently normally distributed, the X's as $N(\mu_1, \sigma_1^2)$ and the Y's as $N(\mu_2, \sigma_2^2)$. We wish to test the null hypothesis H_0, that the two variances are the same, against one of three alternative hypotheses, H_1. The three situations are summarised below:

$$H_0 : \sigma_1^2 = \sigma_2^2 , \qquad \begin{cases} \text{(a)} & H_1 : \sigma_1^2 > \sigma_2^2 \\ \text{(b)} & H_1 : \sigma_1^2 < \sigma_2^2 , \text{ or} \\ \text{(c)} & H_1 : \sigma_1^2 \neq \sigma_2^2 . \end{cases}$$

For this purpose, we use the F test with the ratio of the two sample variances as our test criterion, T. In case (a), we use

$$T = s_1^2/s_2^2 ,$$

which is distributed as F_{ν_1,ν_2} under H_0, where the degrees of freedom $\nu_1 = n_1-1$ and $\nu_2 = n_2-1$ are simply those associated with the two sample variance estimates. We reject H_0 at the significance level α if $t \geq F_{\nu_1,\nu_2}(1-\alpha)$, the $1-\alpha$ quantile of the distribution of F_{ν_1,ν_2}. In case (b), we use

$$T = s_2^2/s_1^2 ,$$

and reject H_0 if $t \geq F_{\nu_2,\nu_1}(1-\alpha)$. Note that cases (a) and (b) are essentially the same since there is nothing special about the suffixes 1 and 2. In case (c), we use

$$T = \frac{\max(s_1^2,s_2^2)}{\min(s_1^2,s_2^2)} ,$$

and reject H_0 at the significance level α if $t \geq F_{\nu_{num},\nu_{den}}(1-\alpha/2)$, where ν_{num} and ν_{den} are the numbers of degrees of freedom associated with the estimates which appear in the numerator and denominator of t, respectively.

We remark that, in the usual table of critical values of F, the number of degrees of freedom relating to the numerator of F appears across the top of the table, while the number of degrees of freedom relating to the denominator is printed down the side. It is sometimes useful to remember that $F_{\nu_1,\nu_2}(\alpha)$ is $1/F_{\nu_2,\nu_1}(1-\alpha)$. Often each page of an F-table relates to a particular significance level α, so that the entries are the $1-\alpha$ quantiles of the F-distributions appropriate to the different pairings of ν_1 and ν_2. Usually ν_1 and ν_2 are listed in F tables in such a way as to assist in harmonic interpolation (see the discussion of the use of a t table in Chapter 6, Section 1). For example, the last few values of ν_1 may be 12, 24, ∞, which are 24/2, 24/1 and 24/0, and the last few values of ν_2 may be 40, 60, 120, ∞, which are 120/3, 120/2, 120/1 and 120/0.

Example. It will be recalled that, in Chapter 6, Section 2 and subsequently in Chapter 8, Section 2, we discussed some analytical data from two different methods, A and B. The essential feature of those discussions was the inequality in the distribution variances of the two methods. Here, we use the observed sample variances to test this assumption.

Formally, our two hypotheses are

$$H_0 : \sigma_A^2 = \sigma_B^2 ,$$

$$H_1 : \sigma_A^2 \neq \sigma_B^2 ,$$

since we have no prior knowledge as to whether one method is likely to be more precise than the other. The sample variances have already been given as

$$s_A^2 = 11.11 \times 10^{-4} \qquad (\nu_A = 9) ,$$

$$s_B^2 = 45.34 \times 10^{-4} \qquad (\nu_B = 9) .$$

Hence the realization of our test criterion is 45.34/11.11 = 4.08, just in excess of the critical value of $F_{9,9}(0.975) = 4.03$ for a two-sided test at the 5 percent significance level. Hence we reject H_0 in favour of H_1 that the two methods are of different precision. Following the advice given in Chapter 7, Section 5, we adopt $s_A^2 = 11.11 \times 10^{-4}$ as the best estimate of the variance associated with method A and $s_B^2 = 45.34 \times 10^{-4}$ as the corresponding quantity associated with method B.

3. Comparison of several variances

We suppose that we have k sets of independent observations represented by $X_{i1}, X_{i2}, \ldots X_{in_i}$, each set distributed as $N(\mu_i, \sigma_i^2)$, i = 1,2,...k. We wish to test between the two hypotheses

$H_0 : \sigma_1^2 = \sigma_2^2 = \ldots = \sigma_k^2$ (i.e. the variances of the k sets are equal),

H_1 : not H_0 (i.e. the variances of the k sets are not all equal).

3.1. Hartley's F_{max} test

This is a quick test which may sometimes be used with advantage. We assume normality and use the test criterion

$$T = \max\{s_i^2\}/\min\{s_i^2\} .$$

Tables of critical values of T are available only for the case of equal numbers of degrees of freedom, (for example, Owen's Handbook of Statistical Tables, 1962). The tabulation applies for up to 12 variances being compared, with 5 and 1 percent significance levels. If the realization of T exceeds the critical value, we reject H_0 at the corresponding significance level. The test is less powerful than those which follow, and tables of critical values are not as readily available as, for example, those for F. Nevertheless, the simplicity of the test makes it worth consideration.

If the appropriate tables of critical values are not available, or if the degrees of freedom are unequal, we may compare our test criterion with $F_{\nu_{num},\nu_{den}}(1-\alpha)$, where ν_{num} and ν_{den} are the degrees of freedom of the numerator and denominator of T. If $t < F_{\nu_{num},\nu_{den}}(1-\alpha)$, we may conclude that our value is non-significant at the α level of significance since the correct critical value always exceeds that for the ordinary F-test of two variances. Such a non-significant result eliminates the need for the special tables and also the computation required for the two tests which follow. However, a significant result here does not necessarily stand.

3.2. Bartlett's test

We have k independent sample variances $s_1^2, s_2^2, \ldots s_k^2$ with degrees of freedom $\nu_1, \nu_2, \ldots \nu_k$; ν_i $(i = 1,2,\ldots k)$ is, of course, equal to $n_i - 1$. We put

$$\nu = \Sigma \nu_i \ ,$$

$$s^2 = \Sigma \nu_i s_i^2 / \nu \ ,$$

and

$$C = 1 + \{(\sum \frac{1}{\nu_i}) - \frac{1}{\nu}\} / \{3(k-1)\} \ .$$

As the test criterion, we use the quantity

$$T = (\nu \ln s^2 - \Sigma \nu_i \ln s_i^2)/C \ .$$

It can be shown that, under H_0, T is distributed as χ^2_{k-1}. Hence we reject H_0 at the significance level α if the realization of T, t, is such that $t \geq \chi^2_{k-1}(1-\alpha)$, the $1-\alpha$ quantile of the distribution of χ^2_{k-1}.

The test is sensitive to non-normality, so that it cannot be relied upon when the normality of the data is seriously in doubt.

3.3. Levene's S test

This test has been shown to be reasonably robust to non-normality when the sample sizes are equal (Levene, 1960). For each set, we define the quantities Y_{ij} by

$$Y_{ij} = (X_{ij} - \bar{X}_i)^2 \ , \ j = 1,2,\ldots n,$$

where $\bar{X}_i = \sum_j X_{ij}/n$, the mean of the ith set. We test whether or not the means of k sets of Y's are significantly different, since these are proportional to the sample variances.

Letting $S_i = \sum_j Y_{ij}$, the sum of the ith set of Y values, $S = \Sigma S_i$, and $m = kn$, our test criterion is

$$T = \frac{\Sigma S_i^2/n - S^2/m}{\sum_i\sum_j Y_{ij}^2 - \Sigma S_i^2/n} \cdot \frac{(m-k)}{(k-1)} .$$

Under H_0, T is approximately distributed as $F_{k-1,m-k}$. Hence we reject H_0, the hypothesis that the variances are equal, at the significance level α if $t \geq F_{k-1,m-k}(1-\alpha)$.

Here, we are performing a one-way analysis of variance on the Y's. One difficulty, however, is that the Y's are not normally distributed, but this is not too important since the analysis of variance test is robust to non-normality. Another difficulty is that the Y values within a set are not independent because each is derived using the same set mean. However, it has been shown that this departure from the proper conditions for one-way analysis of variance causes only a very small disturbance unless the number of observations in the various sets is very small.

Example. The grouped enthalpy data of Table 3.1 to which we have so often referred may once again be used to illustrate these procedures. These data provide estimates of the variances of the 13 groups, as shown in column 8 of that table, each estimate being associated with 4 degrees of freedom since there are 5 data per group.

Applying the F_{max} test first and referring to Table 3.1, we have the test criterion,

$$t = 6.74/0.38 = 17.7,$$

with degrees of freedom, 4,4. The critical value for significance level 5 percent is 20.4, so we have a non-significant result and accept H_0.

Incidentally, the 0.95 quantile for $F_{4,4}$ is 9.12 so that had we used the ordinary F tables rather than those appropriate to the F_{max} test, we should have obtained an apparently significant result. However, as we have explained, apparently significant results indicated by the use of ordinary F tables need verification using the proper tables. This example shows that the apparent significance may not be real.

For Bartlett's test, we have

$$s^2 = 2.895 ,$$

$$\nu = 52 ,$$

$$C = 1.090 ,$$

and $$t = \frac{52 \times 1.063 - 4 \times 10.68}{1.090} = 11.52 .$$

The critical value of $\chi^2_{12}(0.95)$, for a test at the 5 percent significance level, is 21.03. Hence we accept H_0, that the variances of the 13 groups are equal.

Following Chapter 7, Section 5, we adopt the pooled estimate of variance $s^2 = \Sigma\nu_i s_i^2/\nu = 2.90$ as the best estimate of σ^2, the variance common to all groups. This value of s^2 is associated with $\nu = 52$ degrees of freedom. We note that the pooled estimate agrees well with the estimate calculated from the data as a whole, ignoring the division into groups (2.726 with 64 degrees of freedom).

Alternatively, using Levene's test, we form Table 9.1.

We calculate the following quantities:

$$S = \sum_{i}\sum_{j} y_{ij} = 150.57 ,$$

$$\sum_{i}\sum_{j} y_{ij}^2 = 969.0473 ,$$

$$\sum_{i} S_i^2/5 = 500.432 ,$$

$$S^2/65 = 348.790 .$$

Table 9.1. Table of values of $y_{ij} = (x_{ij} - \bar{x}_i)^2$, where the x_{ij} come from Table 3.1

Group	y					Sum
1	0.31	3.03	1.35	0.29	0.31	5.29
2	7.08	1.12	2.37	9.86	0.92	21.35
3	3.31	2.50	0.77	0.00	0.38	6.96
4	1.21	0.25	0.49	0.36	3.61	5.92
5	0.13	0.19	0.71	0.07	0.44	1.54
6	3.84	0.13	2.43	12.53	0.12	19.05
7	0.05	9.49	2.50	3.31	6.86	22.21
8	2.25	0.49	0.81	2.56	2.25	8.36
9	0.21	1.85	2.43	1.80	4.16	10.45
10	1.39	0.38	2.82	0.52	2.31	7.42
11	0.10	0.46	0.01	3.31	2.50	6.38
12	2.76	0.92	0.41	0.03	4.58	8.70
13	5.76	0.49	3.24	16.81	0.64	26.94

We use these to form the ANOVA table, Table 9.2.

Table 9.2. ANOVA table

Source	d.f.	sum of squares	mean square
Between groups	12	151.642	12.6368
Within groups	52	468.615	9.0118
Total	64	620.257	

The test criterion,

$$t = 12.6368/9.0118$$
$$= 1.402 ,$$

with degrees of freedom 12,52 , is nonsignificant, being less than $F_{12,52}(0.95) = 1.94$. We conclude that there is no significant difference between the variances, so we again accept H_0 and use $s^2 = 2.90$ as our estimate of σ^2.

4. Variance stabilization

Often the reason for testing for homogeneity of variance is to see whether or not this assumption is justified, it being required for certain tests on means. When the variances are found to be significantly different, the difficulty can often be overcome by transforming the data so as to stabilize the variance. However, such a transformation is only possible when there is a relationship between the variance and the mean. We should point out that if the original data were normal then the transformed data would not be so. However, for tests on means, including analysis of variance and regression (see Chapters 12 to 16), the assumption of homogeneity of variance is much more crucial than the assumption of normality. The tests are robust to reasonable departures from normality, but are sensitive to heterogeneity of variance. Thus it is reasonable to use such variance-stabilizing transformations.

Let X be one of our original observations with mean μ and variance σ^2, and let Y be the transformed random variable derived from X. We now list some frequently-occurring relationships between σ^2 and μ, with corresponding suitable transformations. In each case c represents a constant.

(i) If $\sigma^2 = c\mu$, put $Y = X^{\frac{1}{2}}$. In the particular case where $X \sim Pn(m)$, then $\sigma^2 = \mu = m$, $c = 1$, and $V(X^{\frac{1}{2}}) \simeq 1/4$. If m is (or observed x_i values are) not large (say less than 10), a better transformation to use is $Y = X^{\frac{1}{2}} + (X+1)^{\frac{1}{2}}$.

(ii) If $\sigma^2 = c\mu^2$, put $Y = \ln X$.

(iii) If $\sigma^2 = c\mu^4$, put $Y = X^{-1}$.

(iv) If $X \sim B(p,n)$, so that $\sigma^2 = np(1-p)$ and $\mu = np$, the standard recommendation is $Y = \arcsin\{(X/n)^{\frac{1}{2}}\}$, but this is unsatisfactory for p near 0 or 1. An improvement is achieved using

$$Y = \arcsin[\{X/(n+1)\}^{\frac{1}{2}}] + \arcsin[\{(X+1)/(n+1)\}^{\frac{1}{2}}] .$$

Further information is given by Freeman and Tukey (1950).

More generally, if $V(X) = g(\mu)$ and we assume that the coefficient of variation of X is small, we may often obtain a suitable transformation as follows. We use the formula for the approximate variance of a function given in Chapter 5, Section 5. Let $Y(X)$ be the function. We require $V(Y) = $ constant, a^2, say. That is,

$$a^2 \simeq \left(\frac{dy}{dx}\right)^2 V(X) \simeq \left(\frac{dy}{dx}\right)^2 g(x) .$$

Hence

$$\frac{dy}{dx} \simeq a/\{g(x)\}^{\frac{1}{2}} ,$$

so that

$$y \simeq \int_b^x \frac{a\,dt}{\{g(t)\}^{\frac{1}{2}}} , \tag{1}$$

where a and b are any convenient constants. Some of the above transformations may be derived by using equation (1).

CHAPTER 10

GOODNESS OF FIT TESTS

By now, it will be clear that prior knowledge or belief plays an important role in the application of statistical reasoning to experimental observations. Sometimes we may be unsure whether or not a particular probability distribution, or a particular _form_ of probability distribution (with some unknown parameters), which we believe to be appropriate for our observations, really is appropriate. In this chapter, we deal with the procedures by which we may test this belief. Essentially, these procedures test just how well our data fit the probability distribution in question by comparing the observed and expected frequencies of occurrence of a series of consecutive sets of possible values.

1. _The Chi-squared test_

There are a number of tests available, but we shall mainly confine ourselves to the chi-squared test of goodness of fit. This test is appropriate for either of the following situations:

(i) when the null hypothesis fully specifies the distribution or

(ii) when the null hypothesis specifies the distribution apart from certain unknown parameters; estimation of these parameters is then required both for the test and for subsequent use of the distribution if the null hypothesis is accepted.

We can summarise the two situations by saying that, under H_0, the distribution function of the population from which the data are drawn is $F(x;\theta)$, where the function F is known, and θ (which may be a vector) may be known or unknown. The alternative hypothesis, H_1, asserts simply that $F(x;\theta)$ _is not_ the distribution function of X, which may be _any other_ distribution function; H_1 is thus a quite general alternative.

For the purposes of the test, we count the numbers of data falling into particular sets of possible values of the observations. In the case of a continuous random variable, these sets are a series of intervals, whereas, if the random variable is discrete, they are either individual values or sets of adjacent values. In either case, the occurrence of one of the sets constitutes an event. Hence, from our data, we obtain a set of observed frequencies or counts of the events.

In the second stage of the test, we calculate the expected frequencies for these events under H_0. By expected frequencies, we mean the total number of data, N say, multiplied by either the probabilities of the events [using $F(x;\theta)$ when the distribution function is fully specified] or by the estimated probabilities of the events (when θ is unknown). In this latter situation, we firstly estimate θ, as θ^* say, by an asymptotically efficient method (usually maximum likelihood) using the observed counts (not the individual observations), and then use $F(x;\theta^*)$ to calculate the estimated probabilities.

For the ith event ($i = 1,2,...k$), let O_i and E_i be the observed and expected frequencies or counts. Then, our test criterion is

$$T = \Sigma(O_i - E_i)^2/E_i \ . \tag{1}$$

For large N and provided that none of the E's is too small under H_0, T is approximately distributed as χ^2_ν where ν, the number of degrees of freedom associated with χ^2, is given by

$$\nu = k - 1 - h \ , \tag{2}$$

k being the number of categories used, and h the number of parameters in the distribution function which have to be estimated. The accepted convention regarding a safe minimum for any E is 5, but in fact one can safely allow a small proportion (a maximum of about one-fifth) of the E's to be a little less than this. In applying the test, we reject

H_0 at the α level of significance if the realization of our test criterion is greater than or equal to $\chi^2_\nu(1-\alpha)$.

For the purposes of calculation, an equivalent and sometimes more convenient form for T is

$$T = \Sigma O_i^2/E_i - N . \tag{3}$$

This is easily derived from equation (1) using the fact that $\Sigma O_i = \Sigma E_i = N$. It should also be noted that when using equations (1) or (3), it is sufficient to work out the E's and the separate contributions to T to two decimal places.

One asymptotically efficient method of estimation of θ, when this is unknown, is the minimum chi-squared method, whereby θ^* is chosen to minimize T. This is asymptotically equivalent to the method of maximum likelihood and could be used here (though it is usually less convenient to apply than the method of maximum likelihood). However, if we were to calculate T in a particular situation using any estimator of θ whatsoever, and we obtained a nonsignificant result, then clearly the correct value of T using minimum chi-squared estimation would be less than or equal to the value we obtained, and so no more significant. Hence nonsignificance stands using any estimator.

Example. Having spent so much time examining the enthalpy of neutralization data of Table 3.1, it is appropriate here to test whether our observations are normally distributed. In this case, we have no prior knowledge of the mean and variance of this distribution so we have to use the data to provide estimates of these quantities.

We proceed as follows. First, we count the data occurring in convenient intervals. Since the maximum and minimum values of x are 61.6 and 53.8 respectively, it is clearly appropriate to use nine intervals of unit width: (53.05, 54.05), (54.05, 55.05), ... (61.05, 62.05). The observed frequencies of occurrence of the data in these intervals are shown in Table 10.1.

Table 10.1. Observed frequencies of occurrence of the x values of Table 3.1

Category	Interval bounds	Organizing the counts	Observed frequency
1	53.05 - 54.05	|	1
2	54.05 - 55.05	||||	4
3	55.05 - 56.05	𝍸 𝍸 |	11
4	56.05 - 57.05	𝍸 𝍸 |	11
5	57.05 - 58.05	𝍸 𝍸 𝍸 ||||	19
6	58.05 - 59.05	𝍸 |||	8
7	59.05 - 60.05	𝍸 |||	8
8	60.05 - 61.05	||	2
9	61.05 - 62.05	|	1

When making these counts, it is useful to use the tally system, entering a bar in the appropriate row or column on each occurrence of a datum until a total of four has been accumulated and then on the next occurrence crossing through the four with a diagonal to give a self-contained group of five. Apart from facilitating the labour of counting, this procedure gives a rough indication of the shape of the corresponding histogram if the bars are entered equally spaced. This is shown in Table 10.1. We note that some of the intervals which we have chosen contain very few observations.

We now estimate the distribution mean and variance from the counts and the mid-points of the intervals exactly as in Chapter 5 where we examined 1000 observations of the transit time of Polaris. Our calculations are set out in Table 10.2, where it will be seen that we have estimated σ^2 by $\hat{\sigma}^2$ (the maximum likelihood estimate) since this is more convenient than s^2; because n = 65, its bias is negligible. We obtain $\hat{\mu} = 57.32$ and $\hat{\sigma}^2 = 2.731$ which are only slightly different from the estimates calculated using the individual observations, 57.34 and 2.760 respectively. We emphasize that the latter estimates are not those appropriate to the test we are about to make.

Table 10.2 Estimation of the distribution mean and variance using the grouped data of Table 10.1

Col: 1	2	3	4	5	6	7	8
x_{mp}	f	y	fy	fy^2	y+1	f(y+1)	$f(y+1)^2$
53.55	1	-4	-4	16	-3	-3	9
54.55	4	-3	-12	36	-2	-8	16
55.55	11	-2	-22	44	-1	-11	11
56.55	11	-1	-11	11	0	-22	0
57.55	19	0	-49	0	1	19	19
58.55	8	+1	8	8	2	16	32
59.55	8	+2	16	32	3	24	72
60.55	2	+3	6	18	4	8	32
61.55	1	+4	4	16	5	5	25
	65		34	181		72	216
			-49			-22	
			-15			50	

Notes

(a) Column 1: the mid-point of the interval, x_{mp}

(b) Column 2: the observed frequency, f

(c) Column 3: the coded variable, $y = x_{mp} - 57.55$

(d) Checks:

(i) $\Sigma f(y+1) = \Sigma fy + \Sigma f$

$= -15 + 65$

$= 50$ ✓

(ii) $\Sigma f(y+1)^2 = \Sigma fy^2 + 2\Sigma fy + \Sigma f$

$= 181 - 30 + 65$

$= 216$ ✓

Table 10.2 (continued)

Estimation

(a) The appropriate estimate of μ is $\bar{x}_{mp}$ $= \bar{y} + 57.55$

$= -15/65 + 57.55$

$= 57.32$

(b) The appropriate estimate of σ^2 is $\hat{\sigma}^2(x_{mp}) = \hat{\sigma}^2(y)$

$= \overline{y^2} - (\bar{y})^2$

$= 181/65 - (-15/65)^2$

$= 2.731$

Finally we use our estimates to calculate the expected frequencies of occurrence of data under the hypothesis H_0 that the observations are normally distributed. However, without performing any computations, it is clear that, if we retain the same intervals, some of our expected frequencies will be less than 5, low values which tend to invalidate the asymptotic χ^2 distribution under H_0. Consequently we try to regroup our data with the objective of getting an expected count of at least five in each interval. Strictly, we should have to employ an iterative process to ensure this, which is both time-consuming and unlikely to affect the outcome of our test. We therefore go for a simpler procedure, regrouping so as to have an observed count of at least five in an interval and hope that our expected counts (when we have calculated them) are not smaller than the recommended value. Table 10.3 shows our new arrangement and that the 'common-sense' regrouping is successful. We now calculate the realization of our test statistic

$$T = \Sigma O_i^2/E_i - N .$$

We obtain $t = 3.5$. Under H_0, this is a realization of χ^2_ν, where

$$\nu = 6 - 1 - 2 = 3 ,$$

(6 groups or categories, 2 estimated distribution parameters). The critical value of $\chi^2_3(0.95)$ for a test at the 5 percent significance level is

Table 10.3 The regrouped data of Table 10.1, and the observed and expected frequencies

New category	Interval bounds LB	UB	$z = \frac{UB-57.32}{2.731}$	$\Phi(z)$	Expected frequency under H_0	Observed frequency	Parts of T
1	$-\infty$	55.05	-1.374	0.0847	5.51	5	0.05
2	55.05	56.05	-0.768	0.2212	8.87	11	0.51
3	56.05	57.05	-0.163	0.4352	13.91	11	0.61
4	57.05	58.05	+0.442	0.6707	15.31	19	0.89
5	58.05	59.05	+1.047	0.8524	11.81	8	1.23
6	59.05	$+\infty$	∞	1	9.59	11	0.21
						Total	3.50

To calculate the expected frequencies, we write for example

$$P(55.05 < X \leq 56.05) = \Phi(-0.768) - \Phi(-1.374) = 0.1365 \ .$$

Hence the expected frequency is $0.1365 \times 65 = 8.8725 \simeq 8.87$.

7.81. Hence we accept H_0, that our data are drawn from a normal population.

When testing the fit of a distribution extending over an infinite range, at least one end interval that we use for our test will be infinite in length, and the estimation of parameters involving the representation of such an interval by a single x value is not feasible. It is a convenient approximation to use intervals that cover all the observations for estimation purposes, then to regroup end intervals to cover the tail(s) and to ensure that the minimum observed count is at least five.

2. Contingency tables

A particular case of a goodness of fit test is the test for independence between two different classifications for the same items. For example, a number of schoolboys may be classified according to which type of school they go to and the socio-economic group of their fathers. Another example might be a group of students classified according to their grades in two subjects.

We apply the chi-squared goodness of fit test to this situation. Our observed counts form a rectangular table, where the rows represent classification according to one factor, and the columns classification according to the other. Each row in the table corresponds to a particular level of the first classification (factor 1), and each column, to a particular level of the second classification (factor 2). Thus the number, O_{ij} say, appearing in the (i,j)th cell of this table is the number of items which are observed at the ith level of factor 1 and the jth level of factor 2.

Under the null hypothesis of independence of the two classifications, the probability p_{ij} of an item falling into the (i,j)th cell is the product of the probability p_i of an item falling into the ith level of factor 1, and the probability q_j of an item falling into the jth level of factor 2. The alternative hypothesis is simply the converse of H_0 so that our two hypotheses are

$$H_0 : p_{ij} = p_i q_j ,$$

$$H_1 : p_{ij} \neq p_i q_j .$$

Now the probability of an item falling into the ith level of factor 1 may be estimated under H_0 from the number of items found altogether in the ith level (the row total, r_i) and the total number of items, thus $\hat{p}_i = r_i/N$. Similarly q_j may be estimated under H_0 from the column total, c_j , thus $\hat{q}_j = c_j/N$.

Here

$$r_i = \sum_j O_{ij} ,$$

$$c_j = \sum_i O_{ij} ,$$

and

$$\sum_i\sum_j O_{ij} = \sum_i r_i = \sum_j c_j = N .$$

Hence we may estimate the expected count in the (i,j)th cell by the simple expression

$$E_{ij} = N(r_i/N)(c_j/N)$$
$$= r_i c_j/N .$$

For the purposes of the test, we calculate

$$T = \sum_i\sum_j (O_{ij}-E_{ij})^2/E_{ij} , \qquad (1)$$

in exact analogy to the test criterion given earlier in equation (1.1). Under H_0, T is approximately distributed as χ^2_ν, where, as before,

$$\nu = k - 1 - h . \qquad (2)$$

The number of distinct categories in our table is simply the total number of cells, rc, where r is the number of rows and c is the number of columns. The number of parameters which we have estimated to give the expected counts is (r-1)+(c-1); this is because we only need (r-1) row probabilities (since the r probabilities add up to 1) and similarly, (c-1) column probabilities. Substituting k = rc and h = (r-1) + (c-1) in equation (2) gives for the number of degrees of freedom associated with χ^2

$$\nu = (r-1)(c-1) . \qquad (3)$$

As in the previous section, we reject H_0 at the α level of significance if $t \geq \chi^2_\nu(1-\alpha)$.

To calculate T from equation (1), we may perhaps form a table showing $|O_{ij} - E_{ij}|$, also E_{ij}, in the (i,j)th cell, then calculate T from this table. Alternatively, we may adopt a similar form to equation (1.3), thus

$$T = N\Big(\sum_i r_i^{-1} \sum_j (O_{ij}^2/c_j) - 1\Big) ,$$

or

$$T = N\Big(\sum_j c_j^{-1} \sum_i (O_{ij}^2/r_i) - 1\Big) .$$

The calculation may be simplified somewhat if at least one of the factors has only two levels, as we shall now show.

2.1. 2 x 2 Table

In this special case, each factor has only two levels. It is convenient to adopt a different notation here, classifying under A or B for one factor and under I or II for the other, and labelling the counts in the four cells a,b,c,d as shown in Table 10.4.

Table 10.4. Table of counts for a 2 x 2 table

	A	B	Total
I	a	b	r
II	c	d	s
Total	m	n	N

It can easily be shown that the usual formula for T [equations (1) or (3)] simplifies to

$$T = \frac{N(ad-bc)^2}{mnrs} , \tag{4}$$

where m,n,r,s are the marginal totals. T is approximately

distributed as χ^2 with 1 degree of freedom under H_0. Notice that the quantity (ad - bc) is the determinant of the 2 x 2 matrix of the table.

There is a correction for continuity ("Yates' correction", or "the half correction") which can be conveniently made to T to improve the approximation for a 2 x 2 table. With this correction, we get

$$T = \frac{N(|ad-bc| - N/2)^2}{mnrs} , \tag{5}$$

so that a smaller T results, which is thus less likely to yield a significant result. However, this correction has been shown (Plackett, 1964) to only produce an improvement in the approximation if all the marginal totals m, n, r, and s are fixed and known in advance before the counts in the columns of the table are observed. In practice this is an unusual situation unless one regards these totals as given, in which case one performs the test conditionally on them using the half correction. However, for most situations which we can envisage it is more appropriate to use equation (4) for the computation of T.

Example. The Joint Matriculation Board's Annual Report for 1974-75 shows the pass rate for boys and girls separately in the Advanced level examination, Pure Mathematics with Statistics. The total number of candidates of each set are also shown. We wish to see whether the pass rate in this examination depends on the sex of the candidates.

Here we have a typical two factor situation (sex, result), each factor with 2 levels (boy, girl; pass, fail). The counts in each cell of the 2 x 2 table are shown in Table 10.5 together with the expected values (shown in brackets in the appropriate cells) estimated from the marginal totals; these latter quantities are presented for interest only since we shall use the special formula (4) for T rather than one of the basic equations (1) or (3). We obtain 18.81 for t, the realization of our test statistic, which is very much greater than $\chi_1^2(0.999) = 10.83$, implying significance at the 0.1 percent significance level. Hence we reject H_0 that the pass rate is independent of the sex of the candidate.

Instead we accept H_1 that the pass rate is influenced by the sex of the candidate and from the data presented in the table conclude that the girls performed significantly better in this examination.

Table 10.5. 2 x 2 table for the Advanced Level results in Pure Mathematics with Statistics (Joint Matriculation Board, Annual Report for 1974-75)

Sex	Result		Totals
	Pass	Fail	
Boy	891 (940.90)	569 (519.10)	1460
Girl	666 (616.10)	290 (339.90)	956
Totals	1557	859	2416

(The counts in the four cells were calculated from the pass rates: 61% for boys and 69.7% for girls, and the total numbers of each sex who entered for the examination).

2.2. 2 x c Table

Here our first factor has two levels, but our second factor c levels. Reverting to our general notation, we see that

$$E_{1j} = c_j(r_1/N) \tag{1}$$

and

$$E_{2j} = c_j - E_{1j} \quad , \tag{2}$$

which makes the calculation simpler, since r_1/N is a constant factor throughout row 1. We may use equations (1) and (2) to form a table of E's and proceed to use equation (2.1) for the evaluation of T.

3. The problem of outliers

An outlier is an observation which does not appear to conform with the rest of the set. For a single univariate sample, an outlier is an extreme observation, rather larger or smaller than the rest of the sample. There may be several such in a sample which do not appear to come from the same probability distribution as the rest of the observations. The problem is to decide whether or not this is so - whether or not there is something faulty about the taking or recording of these far-from-typical observations. We thus have a similar task to that with which we have been concerned in the earlier part of this chapter. Here, we have to examine whether the probability distribution of the data could reasonably give rise to the doubtful observations.

We must warn against too readily or too lightly discarding an observation as an outlier - there is often quite a temptation to do this, particularly if the suspect observation seems to go against a conclusion the rest of the sample appears to be indicating, or perhaps against a conclusion which we would like to arrive at. For this reason, we recommend using a 1 percent significance level rather than the usual 5 percent. The process leading to an outlier being suspected is somewhat subjective and a useful discussion of this is given by Collett and Lewis (1976).

Case 1. Normal data with no further information

Most work on testing for outliers has been done assuming that the normal distribution applies for the observations other than any outliers. A test proposed by Ferguson (1961) uses test criteria

$$C_1 = m_3/m_2^{1.5} \tag{1}$$

and

$$C_2 = m_4/m_2^2 , \tag{2}$$

where the sample moments, m_r , are defined by the expression $m_r = \Sigma(X_i-\bar{X})^r/n, (r \geq 2)$. These quantities are calculated using all the data, including the suspect values. The same test criteria are often used to test whether or not the data conform to the normal distribution. C_1 tests for possible skewness (lack of symmetry), and C_2 for kurtosis (that is, how blunt or sharp the peak is for a given σ^2). So we use

C_1 to test for outliers if the suspect observations lie on one side of the rest of the data, and we use C_2 if the suspect observations lie on both sides. To perform the test, we compare the test criterion with a critical value obtained from tables (e.g. Biometrika Tables for Statisticians Vol.1; C_1 given for $n \geq 25$, C_2 for $n \geq 200$). In the case of C_1 we compare our value with + or - the tabulated value according to whether the suspected outliers lie above or below the rest of the data.

Case 2. Normal data with known σ

In this case we use the test criteria

$$U = (X_{max} - \bar{X})/\sigma \quad \text{or} \quad (\bar{X} - X_{min})/\sigma ,$$

where X_{min} and X_{max} are the smallest and largest observations (the suspected outliers) in the sample of size n. U is called the standarized extreme deviate. Tables of critical values of U are given in the Biometrika Tables.

Case 3. Normal data with an independent estimator s of σ

Here we use the criteria

$$T = (X_{max} - \bar{X})/s \quad \text{or} \quad (\bar{X} - X_{min})/s ,$$

where T is called the studentized extreme deviate. It is important that s is calculated from another sample. Again, critical values for these criteria appear in the Biometrika Tables.

Case 4. Data with fully known distribution

The data are transformed so that the new data, $\{Y_i\}$ say, have the rectangular (0,1) distribution, by putting $Y = F(X)$. If the largest datum is suspect, we use the test criterion

$$Z = \Sigma Y_i / Y_{max} .$$

Under H_0, that the data are independently, rectangularly distributed over (0,1), Z will be approximately distributed as $N\{\frac{1}{2}(n+1),(n-1)/12\}$, so that a normal test may be used. The approximation is satisfactory even for n as low as 3. If the smallest datum is suspect, we may put $W_i = 1-Y_i$, all i, and use $Z = \Sigma W_i/W_{max}$.

Example. We demonstrate the procedure of Case 1 by examining the data of Table 3.1 for skewness. We might, for example, be slightly concerned about the appearance of the value 53.8. We therefore calculate the second and third central moments of our sample as

$$m_2 = 2.68 ,$$

$$m_3 = 0.88 .$$

We compute the test statistic C_1 as 0.20. This is less than the 1 percent critical value 0.698 for n = 65, the number of observations in our sample, and so we conclude that there are no outliers in our data set.

It will be clear from this example that the test is relatively insensitive and that only extreme discrepancies are likely to be detected by it.

Just to illustrate where the test does succeed in revealing an outlier, we use a fictitious example derived from the data of Table 3.1. We replace the value 53.8 by 35.8, which might appear if the first two digits were accidentally transposed. Now

$$m_2 = 9.56 ,$$

$$m_3 = -144.33 ,$$

and

$$C_1 = -4.89 .$$

Our value of C_1 here is far smaller than the 1 percent critical value of -0.698, so we reject the null hypothesis that there is no outlier and now believe the value 35.8 to be inconsistent with the rest of the data.

CHAPTER 11

CORRELATION

To examine the relationship between two random variables X and Y, it is usually advisable to firstly prepare a <u>scatter diagram</u>, which is a plot of the observations $\{x_i, y_i\}$. Such a diagram shows whether

(i) the values of x and y tend to lie approximately on a straight line, or

(ii) the values of x and y tend to lie approximately on some other type of curve, or

(iii) the values of x seem to be quite unrelated to those of y.

Here (iii) is the same as (i) if the slope of the straight line is zero. These three cases are illustrated by Figures 11.1 (i), (ii), and (iii). The purpose of this chapter is to introduce briefly the idea of correlation. We defer detailed consideration of linear and polynomial relationships between X and Y to the next four chapters.

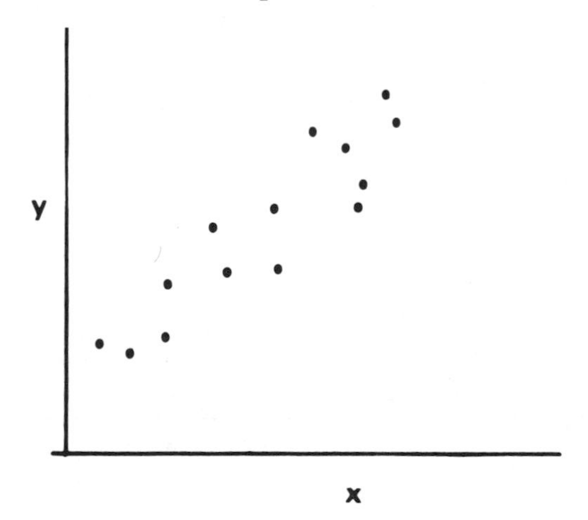

Figure 11.1. (i) A scatter diagram for an approximate linear relationship

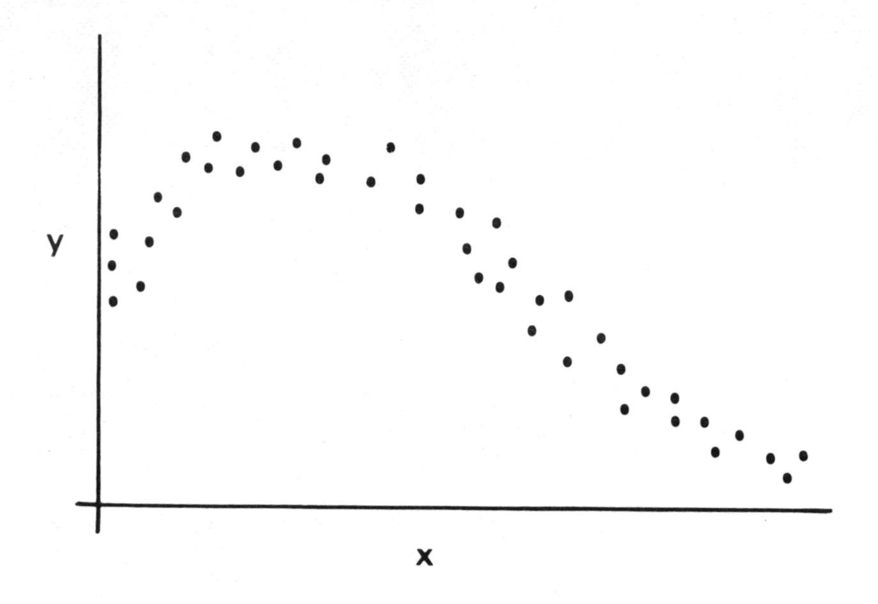

Figure 11.1. (ii) A scatter diagram for an approximate non-linear relationship

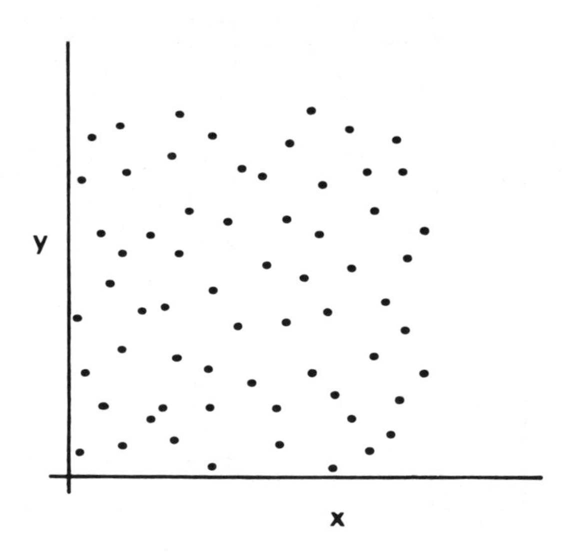

Figure 11.1. (iii) A scatter diagram for no relationship

1. Sample and distribution correlation coefficients

If we have a sample of observations $\{X_i, Y_i\}$, $i = 1,2,\ldots n$ then a measure of the tendency for X and Y to be linearly related is the sample correlation coefficient, r, defined as

$$r = \frac{\Sigma(X_i-\bar{X})(Y_i-\bar{Y})}{\{\Sigma(X_i-\bar{X})^2\Sigma(Y_i-\bar{Y})^2\}^{\frac{1}{2}}} . \qquad (1)$$

A more general form, for weighted data, is given in Chapter 14, Section 3.4. For convenience of computation, we may use the relations

$$\Sigma(X_i-\bar{X})^2 = \Sigma X_i^2 - (\Sigma X_i)^2/n \qquad \text{(similarly for Y)},$$

and

$$\Sigma(X_i-\bar{X})(Y_i-\bar{Y}) = \Sigma X_i Y_i - (\Sigma X_i)(\Sigma Y_i)/n .$$

We have already mentioned in Chapter 3 the corresponding measure of linear relation in the distribution of X and Y. This is the distribution correlation coefficient, ρ, defined by

$$\rho = \frac{C(X,Y)}{\{V(X).V(Y)\}^{\frac{1}{2}}} ,$$

where ρ has the same sign as $C(X,Y)$. In fact an appropriate estimator of ρ is r.

We note that both r and ρ are dimensionless. Further $|r| \leq 1$ and $|\rho| \leq 1$, the equality in both cases corresponding to an exact linear relationship; thus when $|r| = 1$, the points $\{x_i, y_i\}$ lie on a straight line, and when $|\rho| = 1$, X and Y lie on a straight line with probability 1.

Note that the sign of r, and that of the 'slope of the cloud' of (x,y) points in the scatter diagram (more accurately, that of the slope of the fitted regression line of y on x, or of x on y - see Chapter 14) are the same and are given by the sign of $\Sigma(x_i-\bar{x})(y_i-\bar{y})$.

2. The joint normal density of X and Y

When X and Y are jointly normally distributed with means μ_1 and μ_2, variances σ_1^2 and σ_2^2, and correlation coefficient ρ, the joint density f(x,y), is

$$\{4\pi^2\sigma_1^2\sigma_2^2(1-\rho^2)\}^{-\frac{1}{2}}\exp[-a/2] ,$$

where

$$a = \{(x-\mu_1)^2/\sigma_1^2 - 2\rho(x-\mu_1)(y-\mu_2)/(\sigma_1\sigma_2) + (y-\mu_2)^2/\sigma_2^2\}/(1-\rho^2) .$$

We observe that five parameters are necessary to fully define the density. If f(x,y) is plotted against x and y using Cartesian axes, the resulting three-dimensional figure is elliptical in section at given densities, the eccentricity of the ellipse depending on the magnitude of ρ. Thus, when $\rho = 0$, the section is circular whereas when $|\rho| = 1$, the section is a line, the shape of the density contours clearly reflecting the extent of correlation of the two random variables.

We note further that, when $\rho = 0$, the joint density factorizes into the product of $f_1(x)$ and $f_2(y)$,the marginal normal density functions for X and Y. Thus, zero correlation (or equivalently, zero covariance) in the case of two jointly normally distributed random variables does imply independence, as mentioned in Chapter 3.

Finally, it may be shown that the random variable corresponding to minus twice the exponent in the joint density, viz.

$$\{(X-\mu_1)^2/\sigma_1^2 - 2\rho(X-\mu_1)(Y-\mu_2)/(\sigma_1\sigma_2) + (Y-\mu_2)^2/\sigma_2^2\}/(1-\rho^2) ,$$

is distributed as χ_2^2 . We shall defer the proof of this statement until Chapter 16.

2.1. Test of the significance of r

When X and Y are jointly normally distributed, r is the maximum likelihood estimator of ρ. If $\rho = 0$, the distribution of

$$T = r\left(\frac{n-2}{1-r^2}\right)^{\frac{1}{2}}$$

is that of Student's t with n-2 degrees of freedom. Hence T is a convenient criterion for testing $H_0 : \rho = 0$, whether H_1 asserts $\rho > 0$, or $\rho < 0$, or $\rho \neq 0$, using the corresponding one-sided or two-sided t test.

Actually, this test is robust to non-normality, particularly for large n, but even with n as small as 12 we may appeal to the robustness of this test to use it under reasonable departures from normality. When X and Y are jointly normal, testing $H_0 : \rho = 0$ is equivalent to testing for the independence of X and Y, but this is not necessarily true in other cases.

Example. Concrete made from high alumina cement (HAC) undergoes a series of complex chemical changes which in the short term can seriously weaken the material. The extent to which the chemical change has occurred may be estimated by a quantity we designate Y from a sample taken from the concrete by a process which is inconvenient, difficult and destructive. We shall call Y the 'degree of conversion'.

It is therefore desirable to measure some other property of the material which changes as the composition changes, which gives some indication of the strength of the material, and which does not suffer from the disadvantages of sampling. The velocity of high frequency sound through the concrete is such a quantity and we here designate this velocity by X.

An investigation into the strengths of a series of buildings was carried out, in which the two measurements of 'concrete quality' were made on the same structural members. The results are shown in Table 11.1 and plotted in Figure 11.2.

Table 11.1 Results for x = (velocity of sound)/km s^{-1} and y = degree of conversion for a series of high alumina concrete members in a number of different buildings

x	y	x	y
3.44	84	4.22	43
3.48	88	4.23	48
3.85	55	4.23	65
3.86	46	4.24	40
3.93	59	4.24	61
3.93	60	4.25	44
3.95	40	4.25	44
3.95	41	4.25	70
3.95	52	4.29	47
3.98	68	4.29	84
3.99	45	4.30	37
4.02	50	4.30	46
4.04	42	4.31	55
4.04	43	4.31	66
4.04	58	4.32	32
4.05	41	4.32	58
4.05	79	4.32	69
4.06	45	4.33	63
4.07	50	4.34	69
4.07	72	4.36	61
4.08	58	4.38	85
4.09	50	4.40	70
4.11	45	4.41	46
4.11	57	4.43	74
4.12	63	4.47	63
4.13	35	4.48	38
4.13	40	4.50	37
4.15	34	4.51	62
4.15	38	4.54	58
4.15	51	4.59	56
4.18	51	4.71	51
4.18	64	4.74	40
4.19	73	4.93	22
4.20	47	4.97	15

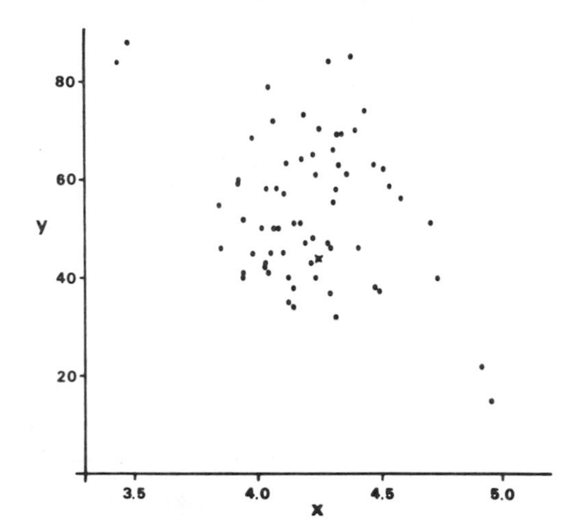

Figure 11.2. Scatter diagram showing y = degree of conversion of HAC plotted against x = sound velocity/km s^{-1}. x = double point.

Because of the complexity of the material, the differences in its preparation on site, and the different environments which it experiences in situ, a high correlation between the two quantities would not be expected. We calculate

$$\Sigma x_i = 3\,644 \qquad \Sigma x_i^2 = 210\,204$$

$$\Sigma y_i = 28\,648 \qquad \Sigma y_i^2 = 12\,115\,774$$

$$\Sigma x_i y_i = 1\,526\,712 \qquad n = 68 .$$

Hence $\Sigma(x_i-\bar{x})^2 = 14\,928.47$,

$$\Sigma(y_i-\bar{y})^2 = 46\,540.12 ,$$

$$\Sigma(x_i-\bar{x})(y_i-\bar{y}) = -8\,483.38 .$$

The sample correlation coefficient is then

$$r = \frac{-8\,483.38}{(14\,928.47 \times 46\,540.12)^{\frac{1}{2}}} = \frac{-8\,483.38}{26\,358.54} = -0.3219 .$$

To examine whether or not this is significant, we set up the two hypotheses

$$H_0 : \rho = 0 ,$$

$$H_1 : \rho \neq 0 ,$$

using a two-sided test since we shall suppose, for simplicity, that we have no prior knowledge of what sign ρ will have if it is not zero. The realization of our test statistic is thus

$$t = -0.3219 \, (66/0.8964)^{\frac{1}{2}} = -2.76 ,$$

which is less than $t_{66}(0.995) = -2.652$. Hence r is significantly large negatively at the 1 percent level, so we reject H_0 and accept H_1 that there is some correlation between the two quantities.

It should be emphasized that the existence of a significant correlation between two random variables does <u>not</u> necessarily imply that one is the <u>cause</u> of the other - there might be one or more other causal factors (such as time) affecting them both. Where a significant correlation occurs between two random variables which have no direct relation to one another, such a correlation is often called a <u>nonsense correlation</u>.

2.2. Linear sums of random normal variables

In Chapter 4, we mentioned that $S = \Sigma a_i X_i$, where $\{X_i\}$ is a set of n random normal variables, is distributed as

$$N\{\Sigma a_i \mu_i \, , \, \sum_{ij}\sum a_i a_j C(X_i, X_j)\} .$$

The simplest case and the one which we discuss here is when $a_i = 1$, all i. Then

$$S \sim N\{\Sigma \mu_i \, , \, \sum_{ij}\sum C(X_i, X_j)\} ,$$

or equivalently

$$\bar{X} = S/n \sim N\{\Sigma \mu_i/n \, , \, \sum_{ij}\sum C(X_i, X_j)/n^2\} .$$

We see from this result that when the X's are positively correlated so that $C(X_i,X_j) > 0$ all i,j, the variance of $\bar{X}$ is greater than it would have been had the X's been independent, with variance unchanged.

Example. A large class of students sits separate examinations in Inorganic, Organic and Physical Chemistry. We assume that their marks are approximately normally distributed with means μ_I, μ_O, μ_P and variances σ_I^2, σ_O^2, σ_P^2, where the subscripts have the obvious meanings. We assume further that there is positive correlation between the marks for the three branches of Chemistry. The average mark for each student in Chemistry is $(X_I + X_O + X_P)/3$, which is normally distributed with

mean $(\mu_I + \mu_O + \mu_P)/3$

and variance $\{\sigma_I^2 + \sigma_O^2 + \sigma_P^2 + 2C(X_I,X_O) + 2C(X_O,X_P) + 2C(X_P,X_I)\}/9$.

To illustrate the effect of the correlations more clearly, we put $\sigma_I^2 = \sigma_O^2 = \sigma_P^2 = \sigma^2$ (possibly achieved by standardizing the marks for the different subjects to cause each set to have a previously chosen mean μ and variance σ^2), then the variance would become

$$\sigma^2(3 + 2\rho_{IO} + 2\rho_{OP} + 2\rho_{PI})/9 .$$

If the correlation coefficients were all one, the variance would become σ^2, the variance of a single mark, whereas if the coefficients were zero the variance would be $\sigma^2/3$. (Incidentally the standardization referred to above could be achieved using

$$\frac{Y_I - \mu}{\sigma} = \frac{X_I - \mu_I}{\sigma_I} ,$$

replacing μ_I and σ_I by the estimates $\bar{x}_I$ and s_I since the number of data is large, and solving for Y_I; similarly for Y_O and Y_P.)

CHAPTER 12

THE STRAIGHT LINE THROUGH THE ORIGIN OR THROUGH SOME OTHER FIXED POINT

In this and subsequent chapters, we shall be concerned with experiments in which two correlated quantities, X and Y, are measured. The outcome of each experiment may thus be represented as a point in a plane. Although originally both X and Y may be random variables, we may be interested in the distribution of Y for given values of X. This may occur in either of two ways:

(i) X and Y may be both random variables, but having observed some values x, of X, we may be interested in considering the conditional distribution of Y for given x, or

(ii) the values of x may be preselected and fixed at certain values (apart from negligible error), not occurring as the result of any random process.

We shall assume throughout this and the next three chapters that the Y values are independent.

In either of these cases the x values are fixed and there is no consideration of any possible underlying probability distribution in the context of this and the subsequent chapters - except to say that (in case (i) above) we could interchange the roles of X and Y.

The distribution of Y for given x has mean $E(Y|x)$, and variance $V(Y|x)$. We shall be concerned with situations where $E(Y|x)$ adopts one of several simple forms, dependent upon x. We shall see that the next four chapters differ from one another in the way in which $E(Y|x)$ is expressed as a function of x. We can make no general remarks about the dependence of the variance of our random variable on x; in some cases we may have reason to assume $V(Y|x)$ to be independent of x, while in others the converse will be true.

This chapter is based on the assumption that

$$E(Y|x) = \beta_1 x , \tag{1}$$

where β_1 is a constant. We add the subscript 1 at this point so as to be consistent with the work of the following chapters. The reasons underlying the choice of equation (1) are most likely to be theoretical in the sense that our understanding of the phenomena involved suggests the direct proportionality of the two quantities. Occasionally, equation (1) may be suggested from the appearance of the scatter diagram obtained by plotting the realizations $\{y_i\}$ of our random variable against $\{x_i\}$, i = 1 to n, where n is the total number of data points.

1. Change of origin

The discussion of this chapter applies equally well to the case where the data are constrained to pass through some fixed point with co-ordinates (x_0, μ_0). All that is necessary here is to shift the origin of the co-ordinate system to the point (x_0, μ_0) by the transformations

$$y' = y - \mu_0 ,$$

$$x' = x - x_0 .$$

The new variable, y' , is then a realization of the random variable Y' whose expectation may be written exactly analogously to the above equation for $E(Y|x)$, namely

$$E(Y'|x) = \beta_1 x' \tag{1}$$

Examples 1 and 2 illustrate the two situations.

Example 1. The elevation, ΔT, of the boiling point of a dilute solution of an involatile, low-molecular-weight solute is expected to be directly proportional to the mass concentration, c, of the solute according to the most general theories of solutions. Hence, at small c the data obtained in the studies of this phenomenon should obey a relation of the type

$$E(\Delta T|c) = \beta_1 c .$$

Example 2. The pressure-volume product Pv for n moles of a real gas at a constant thermodynamic temperature T is expected to depend linearly on 1/v at low pressures, thus

$$Pv = nRT + B/v ,$$

where B is a constant for the particular gas. This equation is analogous to equation (1) since the quantity nRT would be known in any experiment designed to measure B. If nRT is subtracted from the realizations of the random variable Pv, the resulting data have expectations directly proportional to 1/v, thus

$$E\{(Pv - nRT)|v\} = B/v .$$

The programme of work which we have set ourselves in this chapter is as follows. First, we consider how we might estimate such quantities as β_1 and $\mu = E(Y|x)$ from a limited set of (x,y) data. Second, we shall discuss what inferences might be drawn from the results of our analysis assuming that the random variables $\{Y_i\}$ are independently normally distributed; for example, we shall show how a confidence interval for β_1 may be calculated. Finally, we shall describe the means by which a number of simple hypotheses may be tested. We shall go through the early part of this programme twice, once for the case where $V(Y|x)$ is independent of x and once for the case where $V(Y|x)$ is proportional to a known function of x.

2. Some remarks on notation

It is convenient at this stage to explain the main features of the notation which we shall use in this and subsequent chapters.

First we shall continue our practice of using Greek characters for parameters and usually corresponding Latin characters for estimators or estimates of them. For example, we shall denote a particular estimator of β_1 by B_1 and its realization by b_1. In general we shall represent our experimental data by $(x_1,y_1),(x_2,y_2),\ldots(x_n,y_n)$, or simply by $\{x_i,y_i\}$. Each y_i is a realization of the random variable Y_i at x_i, which we may refer to simply as Y_i. We shall write μ_i for the expectation of this random variable and μ for the expectation of Y at a general x within the range where our model applies, thus

$$E(Y_i) = E(Y_i|x_i) = \mu_i ,$$

$$E(Y) = E(Y|x) = \mu ,$$

the population means of Y_i and Y. We shall designate particular estimators of the set $\{\mu_i\}$ by $\{\hat{\mu}_i\}$ and of μ by $\hat{\mu}$. For most of the subsequent work we shall deal with estimators, rather than estimates. We ought to mention at this point that many other text books on this subject use $\hat{y}_i$ where we use $\hat{\mu}_i$, or $\hat{y}$ where we use $\hat{\mu}$, but we are really estimating the <u>parameters</u> μ_i and μ, so our notation is preferable.

3. Formal statement of the first model

We shall assume that there exists a simple relation between the random variable Y and the variable x represented by the equation

$$Y = \beta_1 x + \varepsilon . \tag{1}$$

Here β_1 is a constant and ε is another random variable which we shall refer to as the error term. Our second assumption refers to this random variable, ε. We shall assume that

$$\left.\begin{aligned} E(\varepsilon) &= 0 \\ V(\varepsilon) &= \sigma^2 \end{aligned}\right\} \text{ for all } x , \tag{2}$$

where σ^2 is an unknown constant. From equations (1) and (2) it follows that

$$\mu = E(Y) = \beta_1 x , \tag{3}$$

$$V(Y) = \sigma^2 , \tag{4}$$

so that our assumption of the constant variance of ε implies that the variance of the variable Y is independent of x.

4. Estimation using the method of least squares

Of the methods outlined in Chapter 5, the method of least squares is particularly suitable for the present problem, because the expectation to be estimated is a known linear function of the unknown quantity β_1 , thus

$$\mu = E(Y) = \beta_1 x . \tag{1}$$

Furthermore, under the additional assumption of normality, the method is efficient, as discussed in Chapter 5. The word regression is usually associated with the least squares estimation of a linear combination of a set of unknown parameters (such as β_1 here). According to this method, we choose an estimator B_1 of β_1 , such that the sum of squares $\$$, defined by

$$\$ = \Sigma(Y_i - \beta_1 x_i)^2 , \tag{2}$$

is a minimum. We now apply this method to our present situation. We put

$$\frac{\partial \$}{\partial \beta} = 0 . \tag{3}$$

Replacing β_1 in the resulting equation by its estimator B_1 gives

$$-2\Sigma x_i(Y_i - B_1 x_i) = 0 ,$$

and hence

$$B_1 = \Sigma x_i Y_i / \Sigma x_i^2 . \tag{4}$$

We note that $\partial^2\$/\partial\beta_1^2 = 2\Sigma x_i^2 > 0$, so that B_1 given by equation (4) has minimised \$.

Now B_1 is a linear combination of the Y's with mean

$$E(B_1) = E(\Sigma x_i Y_i/\Sigma x_i^2) = \Sigma x_i E(Y_i)/\Sigma x_i^2 = \Sigma x_i \beta_1 x_i/\Sigma x_i^2 = \beta_1 , \quad (5)$$

so that B_1 is an unbiased estimator of β_1. Since the Y's are independent, its variance is given by

$$V(B_1) = \Sigma(x_i/\Sigma x_i^2)^2 V(Y_i) = \sigma^2/\Sigma x_i^2 . \quad (6)$$

Since $\mu = \beta_1 x$, we estimate μ by

$$\hat{\mu} = B_1 x , \quad (7)$$

which has mean

$$E(\hat{\mu}) = \beta_1 x = \mu , \quad (8)$$

and variance

$$V(\hat{\mu}) = x^2 V(B_1) = x^2\sigma^2/\Sigma x_i^2 . \quad (9)$$

We note that $\hat{\mu}$ is also unbiased and that $V(\hat{\mu})$ increases with x , in contrast to the constancy of $V(Y)$.

4.1. Dependence of $V(B_1)$ on x and n

It is of interest to consider the factors which determine the value of $V(B_1)$. Looking at equation (4.6), we see that increasing the precision of Y, that is, reducing its variance, σ^2, increases the precision of the slope, that is reduces $V(B_1)$. We also see that the larger the term Σx_i^2 , the greater the precision of the slope. This second factor is related to the range covered by the experimental values of x and to their number; large values of $|x|$ will contribute substantially to this sum; increasing numbers of such values will also increase the sum.

We consider the special case of a set of equally-spaced x values, such that

$$x_i = ik , \quad i = 1 \text{ to } n .$$

Then

$$\Sigma x_i^2 = k^2 \sum_{i=1}^{n} i^2 ,$$

$$= k^2 n(n+1)(2n+1)/6 ,$$

$$= x_n^2 (2n + 3 + 1/n)/6 .$$

Hence

$$V(B_1) = \frac{6\sigma^2}{x_n^2 (2n + 3 + 1/n)} . \qquad (1)$$

We conclude, for example, that doubling the maximum value of x while keeping the number of observations constant results in a four-fold decrease in $V(B_1)$; on the other hand, doubling the number of points in a given range of x reduces $V(B_1)$ by somewhat less than a factor of two.

This analysis is relevant to the experimental situation even when the x values are unequally spaced. Provided that the x values are roughly equally spaced over the whole range of x, we may expect that the above conclusions will apply more or less.

If all the n points were at x_n, the variance would be

$$V(B_1) = \sigma^2/(n x_n^2) , \qquad (2)$$

which is clearly smaller again than the value in equation (1). However, such an arrangement of points gives no visual or other confirmation of the model. If such confirmation is not needed, clearly putting all the points at the maximum possible $|x|$ affords the most efficient design, from the point of view of minimising $V(B_1)$.

In Chapter 5, we said that it was desirable for an estimator to have a high probability of being near to the parameter it was estimating. Also when choosing from a set of available unbiased estimators, we should choose that one which has the minimum variance. We are now pointing out that for a given estimator, we may sometimes alter the design of the experiment so as to reduce the variance of the estimator. For the least squares estimator B_1, this can be achieved by:

(a) reducing V(Y) , perhaps by increasing the precision of measurement;

(b) maximising Σx_i^2 by making the largest x value as big as possible (though not outside the range of validity of the model), by having the other x values large too, and by making n as large as possible.

4.2. The estimator, S^2, of σ^2 and derived variance estimators

In most applications of this work, we need to go considerably further than the calculation of the estimators, B_1 and $\hat{\mu}$. The usual practice is to postulate some particular distribution of the Y's and thence to make inferences from the data. First of all we must estimate the quantity σ^2, which appears in the formulation of our model, and in the variances of the estimators B_1 and $\hat{\mu}$ [equations (3.2), (4.6) and (4.9)]. Without appeal to the distribution of the Y's, at this stage, we estimate σ^2 by the least squares estimator which we call S^2. We shall refer to

$$D_i = Y_i - \hat{\mu}_i = Y_i - B_1 x_i$$

as the ith residual (though we may also use the same term for the realization, d_i, of D_i). We use our minimised sum of squares, $\$_{min}$ (see Section 4), which is the sum of squared residuals,

$$\$_{min} = \Sigma D_i^2 = \Sigma(Y_i - \hat{\mu}_i)^2 ,$$

to derive our estimator S^2. ΣD_i^2 is also sometimes called the sum of squares about the regression.

Now

$$\Sigma D_i^2 = \Sigma(Y_i - \hat{\mu}_i)^2 = \Sigma Y_i^2 - 2\Sigma Y_i \hat{\mu}_i + \Sigma \hat{\mu}_i^2 . \quad (1)$$

Also

$$\Sigma Y_i \hat{\mu}_i = \Sigma Y_i x_i (\Sigma Y_i x_i / \Sigma x_i^2) = (\Sigma Y_i x_i)^2 / \Sigma x_i^2 ,$$

and

$$\Sigma \hat{\mu}_i^2 = B_1^2 \Sigma x_i^2 = (\Sigma Y_i x_i / \Sigma x_i^2)^2 \Sigma x_i^2 = (\Sigma Y_i x_i)^2 / \Sigma x_i^2 = \Sigma Y_i \hat{\mu}_i .$$

Hence

$$\Sigma D_i^2 = \Sigma Y_i^2 - \Sigma \hat{\mu}_i^2 , \quad (2)$$

$$= \Sigma Y_i^2 - (\Sigma Y_i x_i)^2 / \Sigma x_i^2 . \quad (3)$$

Here equation (3) is more convenient for computation. This exemplifies the typical result that <u>the sum of squares about the regression</u> (here ΣD_i^2) <u>equals the original sum of squares</u> (here ΣY_i^2) <u>minus the sum of squares due to the regression</u> [here $(\Sigma Y_i x_i)^2 / \Sigma x_i^2$]. Indeed this relation defines the sum of squares due to the regression. Taking expectations of both sides of equation (2), and using the facts that

$$V(Y_i) = E(Y_i^2) - \mu_i^2 ,$$

$$V(\hat{\mu}_i) = E(\hat{\mu}_i^2) - \mu_i^2 ,$$

we obtain

$$E(\Sigma D_i^2) = \Sigma V(Y_i) - \Sigma V(\hat{\mu}_i) ,$$

$$= n\sigma^2 - \Sigma \sigma^2 x_i^2 / \Sigma x_i^2 ,$$

$$= (n-1)\sigma^2 , \quad (4)$$

using equations (3.4) and (4.8). This exemplifies the usual situation to be seen in subsequent chapters, that when we need m estimators to form the estimator $\hat{\mu}$ of μ, the expectation of the sum of squared residuals is reduced by $m\sigma^2$ from $n\sigma^2$ to $(n-m)\sigma^2$. Moreover, the number of degrees of freedom of this sum of squares (and proportional quantities) is n-m, the number of points minus the number of estimated parameters.

From equation (4), it follows that an unbiased estimator of σ^2 is given by

$$s^2 = \Sigma D_i^2/(n-1) , \tag{5}$$

$$= \{\Sigma Y_i^2 - (\Sigma Y_i x_i)^2/\Sigma x_i^2\}/(n-1) , \tag{6}$$

the latter being the form most convenient for calculation purposes. Here s^2 has n-1 degrees of freedom.

Clearly, since

$$V(B_1) = \sigma^2/\Sigma x_i^2 ,$$

$$V(\hat{\mu}) = x^2\sigma^2/\Sigma x_i^2 ,$$

appropriate estimators of these variances, obtained by replacing σ^2 by s^2, are

$$s^2(B_1) = s^2/\Sigma x_i^2 , \tag{7}$$

$$s^2(\hat{\mu}) = x^2 s^2/\Sigma x_i^2 , \tag{8}$$

in both cases with n-1 degrees of freedom, the number associated with s^2. These results apply whether or not x corresponds to a datum value, x_i. We see that $s(\hat{\mu})$ increases linearly with x, though the coefficient of variation, $s(\hat{\mu})/\hat{\mu}$, remains constant. This is shown in Figure 12.1.

We remark that many research workers present their results in graphical form with 'error-bars' drawn in as a visual indication of error. The most common convention is to draw the error-bar on a data point, x_i, Y_i, such that it lies parallel to the Y-axis with length twice the estimated standard deviation of Y_i (in the present context 2s) centred on the data point in question. For a fitted value, a corresponding error-bar would have a length $2s(\hat{\mu})$. However, since there is no universally agreed convention as to what error-bars should stand for, we shall not include them on our graphs.

It is more usual for a statistician to show a confidence band about a fitted line. This is such that the upper and lower bounds corresponding to any x afford a confidence interval for μ at that x. The derivation of such an interval is given later in Section 8.

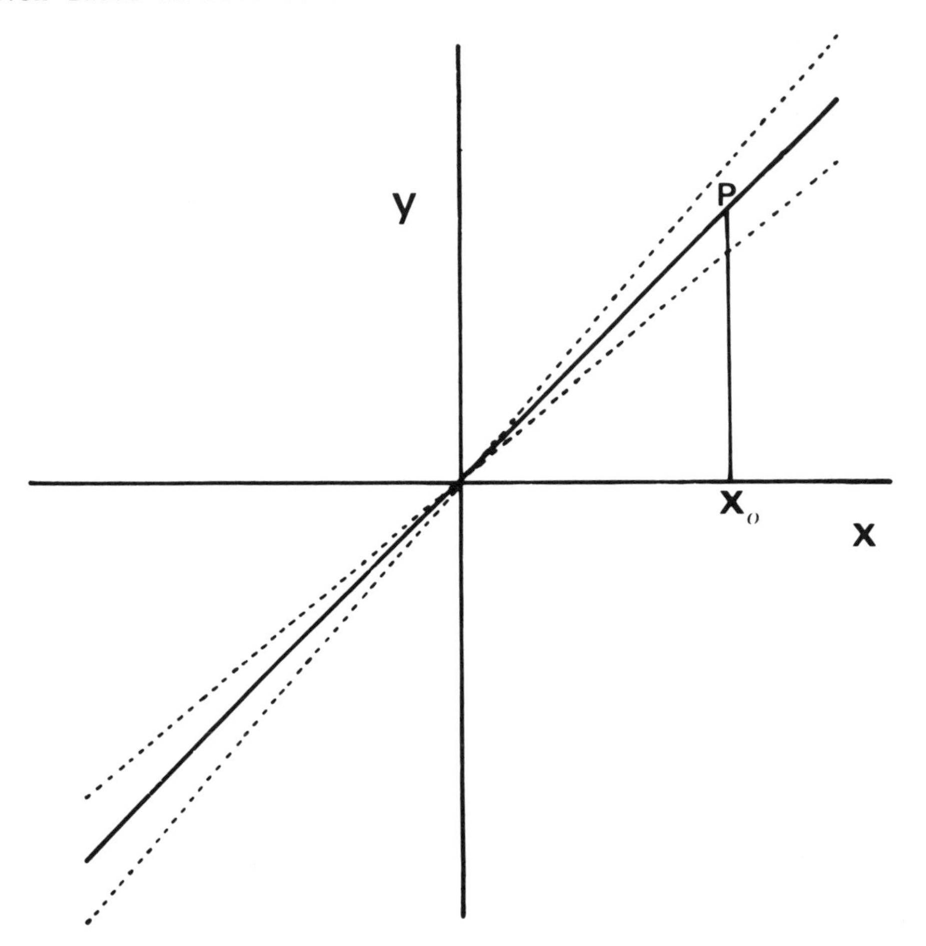

Figure 12.1 The solid line is the least squares line of slope b_1 passing through the origin and the set of points $\{x_i, \hat{\mu}_i\}$. The co-ordinates of the point P are, therefore, (x_0, b_1x_0). The dotted lines show the least squares estimates displaced vertically by ± one standard error, $s(\hat{\mu})$. The diagram shows clearly that the uncertainty associated with a least squares estimate of μ increases rapidly with increasing displacement of x_0 from the origin.

4.3. Inverse interpolation

Occasionally the following type of problem may arise. We perform a set of experiments in which we observe values of Y at preselected x values. From these data we calculate the various parameters already described, namely B_1 and thence $\hat{\mu}$ given by

$$\hat{\mu} = B_1 x . \qquad (1)$$

We now make a further independent observation on Y, for which we wish to estimate the corresponding value of x. Clearly from equation (1) a suitable estimator is

$$\hat{x}_Y = Y/B_1 , \qquad (2)$$

so that $(\hat{x}_Y, Y)$ lies on the least squares line for the original data.

We may derive the approximate mean and variance of $\hat{x}_Y$, using equation (5.5) of Chapter 5. Approximately, $E(\hat{x}_Y)$ is given by $E(Y)/E(B_1) = \beta_1 x/\beta_1 = x$, so that $\hat{x}_Y$ is approximately unbiased. The estimated variance of $\hat{x}_Y$ is

$$S^2(\hat{x}_Y) = \frac{S^2}{B_1^2}\left(1 + \frac{\hat{x}_Y^2}{\Sigma x_i^2}\right) . \qquad (3)$$

If we wish to estimate x corresponding to a given value of μ, we take

$$\hat{x}_\mu = \mu/B_1 , \qquad (4)$$

which again is approximately unbiased with variance estimator given by

$$S^2(\hat{x}_\mu) = \frac{S^2\mu^2}{B_1^4 \Sigma x_i^2} . \qquad (5)$$

5. Formal statement of the second model

As before we shall assume a simple relation between Y and x of the form

$$Y = \beta_1 x + \varepsilon , \tag{1}$$

where β_1 is an unknown constant and ε is the error term. We now assume that

$$E(\varepsilon) = 0 , \quad \text{for all } x, \tag{2}$$

and

$$V(\varepsilon) = \sigma_i^2 = \sigma^2/w_i , \quad \text{for } x = x_i , \tag{3}$$

where σ^2 is assumed to be unknown, and the w's, which we shall call weights or weighting factors, are known for all i. Thus, in this alternative model, the variance of ε , and so of Y, varies over the range of x. From equations (1), (2) and (3), it follows that

$$E(Y) = \beta_1 x , \tag{4}$$

$$V(Y_i) = \sigma_i^2 = \sigma^2/w_i \quad \text{for } x = x_i . \tag{5}$$

If the w_i were all unknown, they could absorb σ^2, and we would effectively have n + 1 unknowns, the w's and β_1, in which case we would be unable to estimate them with only n observed Y values. As it is, with $\{w_i\}$ known, β_1 and σ^2 are the only unknowns. We observe that we could multiply all the w's by any arbitrary constant the inverse of which could be absorbed into the unknown σ^2. We see from equation (5) that the larger w_i is, the smaller is the variance of Y_i.

5.1. An example of the need for weighting factors

Very often, replicate determinations of Y are made at each value of x_i, but the number of replicates varies from one x_i value to another. If Y satisfies the model

$$E(Y) = \beta_1 x , \tag{1}$$

and

$$V(Y) = \sigma^2$$

and if we decide to use the means of the replicate determinations of Y at each value of x_i in our computation of the estimates of β_1 etc. (possibly only the means will be available to us), then weighting factors proportional to the numbers of replicate determinations are required. The formal arguments are as follows. Suppose that n_i observations, Y_{ij}, $j = 1$ to n_i, are made on the random variable Y at $x = x_i$. From these we calculate

$$\bar{Y}_i = \sum_j Y_{ij}/n_i , \tag{2}$$

the mean of the replicate determinations at x_i. The expectation and variance of $\bar{Y}_i$ are

$$E(\bar{Y}_i) = \beta_1 x_i , \tag{3}$$

$$V(\bar{Y}_i) = \sigma^2/n_i , \tag{4}$$

as follows from Chapter 3, Section 4. Comparing equation (4) with equation (5.5), we see that

$$w_i = n_i . \tag{5}$$

Equation (5) shows that if we propose to compute B_1, the estimator of β_1, from a set of data: $(x_1,\bar{Y}_1)$, $(x_2,\bar{Y}_2)$, ... $(x_i,\bar{Y}_i)$, ... $(x_{ir},\bar{Y}_r)$, then those points in which the means are based on the most replicate determinations will be weighted the most heavily.

Another situation in which we can derive appropriate weights for the Y's is when the variance of Y is a known function of x. A good example is provided in Example 2 of Chapter 14.

6. Estimation using the method of least squares

In this more general case, we choose our estimator of β_1 so as to minimise the weighted sum of squares

$$\$ \quad = \quad \Sigma w_i(Y_i-\mu_i)^2 \ .$$

This method is equivalent to the method of maximum likelihood when the data are normal. Reasoning as before, using the derivatives of \$, we can easily see that the minimum of \$ is obtained when

$$B_1 \quad = \quad \Sigma w_i x_i Y_i/\Sigma w_i x_i^2 \ . \tag{1}$$

We thus conclude that this is the least squares estimator of β_1, with expectation β_1 for precisely the same reason as discussed earlier. It is worthwhile evaluating the variance of B_1 to show how the presence of weighting factors modifies the argument of Section (4.2). From equation (1), because the Y's are independent, we obtain

$$V(B_1) \quad = \quad \Sigma w_i^2 x_i^2 V(Y_i)/(\Sigma w_i x_i^2)^2 \ . \tag{2}$$

Using equation (5.5), this becomes

$$V(B_1) \quad = \quad \sigma^2/\Sigma w_i x_i^2 \ . \tag{3}$$

Precisely as before

$$\hat{\mu} \quad = \quad B_1 x \ . \tag{4}$$

$$E(\hat{\mu}) \quad = \quad \mu \ , \tag{5}$$

$$V(\hat{\mu}) \quad = \quad x^2\sigma^2/\Sigma w_i x_i^2 \ . \tag{6}$$

6.1. The estimator, S^2, of σ^2 and derived variance estimators

We proceed as in Section (4.2) to derive the expectation of the random variable $\Sigma w_i D_i^2 = \Sigma w_i(Y_i-\hat{\mu}_i)^2$. We have

$$\Sigma w_i D_i^2 \quad = \quad \Sigma w_i Y_i^2 - 2\Sigma w_i Y_i \hat{\mu}_i + \Sigma w_i \hat{\mu}_i^2 \ . \tag{1}$$

Also

$$\Sigma w_i Y_i \hat{\mu}_i \quad = \quad \Sigma w_i Y_i x_i(\Sigma w_i Y_i x_i/\Sigma w_i x_i^2)$$

$$= \quad (\Sigma w_i Y_i x_i)^2/\Sigma w_i x_i^2 \ ,$$

and

$$\Sigma w_i \hat{\mu}_i^2 = \Sigma w_i x_i^2 (\Sigma w_i Y_i x_i / \Sigma w_i x_i^2)^2$$

$$= (\Sigma w_i Y_i x_i)^2 / \Sigma w_i x_i^2$$

$$= \Sigma w_i Y_i \hat{\mu}_i \quad .$$

Hence

$$\Sigma w_i D_i^2 = \Sigma w_i Y_i^2 - \Sigma w_i \hat{\mu}_i^2 \tag{2}$$

$$= \Sigma w_i Y_i^2 - (\Sigma w_i Y_i x_i)^2 / \Sigma w_i x_i^2 \ .$$

Again this illustrates the relation: the sum of squares about the regression equals the original sum of squares minus the sum of squares due to regression.

Equation (2) here is analogous to equation (4.2.2). It follows just as in Section (4.2) that

$$E(\Sigma w_i D_i^2) = (n-1)\sigma^2 \ ,$$

and that

$$S^2 = \Sigma w_i D_i^2 / (n-1) \tag{3}$$

is an unbiased estimator of σ^2.

As in Section (4.2), we replace σ^2 in the variances of B_1 and $\hat{\mu}$ given by equations (6.3) and (6.6) by S^2 to obtain appropriate estimators of $V(B_1)$ and $V(\hat{\mu})$ as

$$S^2(B_1) = S^2 / \Sigma w_i x_i^2 \ , \tag{4}$$

$$S^2(\hat{\mu}) = x^2 S^2 / \Sigma w_i x_i^2 \ . \tag{5}$$

Each of the variance estimators, S^2, $S^2(B_1)$ and $S^2(\hat{\mu})$ has n-1 degrees of freedom.

6.1.1. A digression re different variance estimators

Example 1 of Section 5.1 considered a set of points $\{x_i, \bar{Y}_i\}$, where $\bar{Y}_i$ is the sample mean of n_i observations, $Y_{i1}, Y_{i2}, \ldots Y_{in_i}$ all at x_i, $i = 1,2,\ldots r$. Put $n = \Sigma n_i$. We could either use all n Y values (each with variance σ^2), if these are available, to estimate β_1, σ^2, etc., or use the r weighted $\bar{Y}_i$ values (with variances σ^2/n_i). The two estimators of β_1 are the same, being

$$B_1 = \sum_{ij}\sum x_i Y_{ij}/\Sigma n_i x_i^2$$
$$= \Sigma n_i x_i \bar{Y}_i/\Sigma n_i x_i^2 .$$

This is a strongly desirable, if not <u>necessary</u>, consistency. However, the two different approaches yield different estimators of σ^2, since the second approach, using weighted means, takes no account of the variability within sets of replicates at the same x values, while the first does. The two sums of squared residuals are

$$\sum_{ij}\sum (Y_{ij} - B_1 x_i)^2 \quad \text{and} \quad \Sigma n_i(\bar{Y}_i - B_1 x_i)^2 ,$$

with n-1 and r-1 degrees of freedom, respectively.

The difference between these two sums of squares,

$$\sum_{ij}\sum (Y_{ij} - \bar{Y}_i)^2 ,$$

is the sum of squares of the deviations of the replicates corresponding to x_i, about their mean $\bar{Y}_i$, summed over all i. It has n-1-(r-1) = n-r degrees of freedom. This is useful for testing the goodness of fit of the data to the model, (see Section 9.3) when this is suspect. All three sums of squares, when divided by their degrees of freedom, provide unbiased estimators of σ^2, under the assumed model.

6.2. <u>Inverse interpolation</u>

Exactly as in Section 4.3, if we wish to estimate the x value corresponding to an independent observation, Y with weight w, we may conveniently use the estimator

$$\hat{x}_Y = Y/B_1 . \qquad (1)$$

As before, this is approximately unbiased, with variance estimator

$$s^2(\hat{x}_Y) = \frac{s^2}{B_1^2}\left(\frac{1}{w} + \frac{\hat{x}_Y^2}{\Sigma w_i x_i^2}\right). \tag{2}$$

Similarly, if a given value of μ is used, $\hat{x}_\mu = \mu/B_1$ and the variance estimator reduces to

$$s^2(\hat{x}_\mu) = \frac{s^2\mu^2}{B_1^4 \Sigma w_i x_i^2}. \tag{3}$$

7. Sampling distributions assuming the normality of the data

In most experimental situations, estimation is not an end in itself - we normally wish to use the estimates to draw further inferences from the data. Exactly what we can do in this connection depends on the form of the probability distributions of the various estimators, which is determined by the type of distribution of Y. Often it will be necessary either to find out experimentally what type of distribution Y has or to assume that a particular type applies. The former of these courses of action is seldom practicable, because of the large number of observations which would be necessary. It is therefore more usual to postulate the form of the distribution of Y. In Chapter 4, we have listed some good reasons why the normal distribution is often chosen though we have warned against making this choice too lightly. These reasons are cogent enough for us to assume in the remainder of this chapter that each Y_i is distributed as $N(\mu_i, \sigma_i^2)$ or, equivalently, each ε_i as $N(0, \sigma_i^2)$, particularly in view of the robustness of the present methods to non-normality. We shall not deal explicitly with the case where the variances of the Y's are constant since the results for this case can always be obtained from the results derived here by putting $w_i = 1$ for all i.

7.1. The distributions of B_1, $\hat{\mu}$, and $\Sigma w_i D_i^2/\sigma^2$

According to equations (6.1) and (6.4), B_1 and $\hat{\mu}$ are linear combinations of the Y's. In Chapter 4, we stated that any linear combination of random variables which are jointly normally distributed is also normally distributed. From this, we conclude that B_1 and $\hat{\mu}$ are normally distributed.

It can be shown (see Chapter 16, Section 2) that $\Sigma w_i D_i^2/\sigma^2$ is distributed as χ^2_{n-1} independently of B_1.

7.2. Important consequences of the assumption of normality

As mentioned earlier when the data are normal, the method of least squares becomes equivalent to the method of maximum likelihood, so that we may appeal to the favourable properties of the latter method.

The estimators B_1 and $\hat{\mu}$ are MVU, and S^2 is unbiased and asymptotically MVU (note the divisor n-1 for S^2 rather than for n as for maximum likelihood does not prevent this property applying). B_1 and S^2 are also jointly sufficient.

8. Confidence intervals for β_1 and μ

The procedure we use here is similar to that described under Case 3 of Section 1 in Chapter 6.

Since $(n-1)S^2(B_1)/V(B_1)$, which equals $\Sigma w_i D_i^2/\sigma^2$ and $(B_1-\beta_1)/\sigma(B_1)$ are independently distributed as χ^2_{n-1} and $N(0,1)$, respectively, the ratio

$$\frac{B_1 - \beta_1}{S(B_1)}$$

is distributed as Student's t with n-1 degrees of freedom. Thus a confidence interval for β_1 is

$$b_1 \pm t_{n-1}(1-\alpha/2)s(B_1) \ ,$$

with confidence coefficient $1-\alpha$.

Similarly, using the fact that $(\hat{\mu}-\mu)/\sigma(\hat{\mu})$ is distributed as N(0,1) independently of S^2, we have $(\hat{\mu}-\mu)/S(\hat{\mu})$ distributed as t_{n-1}. Thus a confidence interval for μ with coefficient $1-\alpha$ is $\hat{\mu} \pm t_{n-1}(1-\alpha/2)s(\hat{\mu})$. For $x = x_i$, we could use $y_i \pm t_{n-1}(1-\alpha/2)s/w_i^{\frac{1}{2}}$ as an interval for μ_i, but this is wider than the interval just stated.

9. Tests of a number of hypotheses

We may wish to test whether or not estimates of the slopes β_1 or the means μ for different lines are reasonably in agreement, or whether or not the slope or mean for a single line is in agreement with a corresponding theoretical value, or possibly literature value, assumed error-free. Alternatively, we may wish to test the validity of the assumed model. We here present standard tests for these purposes.

9.1. Comparison of an 'experimental' slope with a theoretical value

We suppose we require to test whether or not the estimate b_1, which we may refer to as the experimental slope, is reasonably in agreement with a theoretical value, $\beta_{1,0}$, for β_1. For a two-sided test we have

$$H_0 : \beta_1 = \beta_{1,0} \quad ,$$

$$H_1 : \beta_1 \neq \beta_{1,0} \quad .$$

Under H_0 our test ratio, $T = (B_1 - \beta_{1,0})/S(B_1)$, will be distributed as t_{n-1}. For our two-sided test with significance level α, we reject H_0 if $|t| \geq t_{n-1}(1-\alpha/2)$. This procedure is identical to that of a two-sided test of a mean in Case 3 of Chapter 8. Corresponding tests apply for the two one-sided situations.

9.2. Comparison of an 'experimental' with a theoretical mean at a particular x

This is exactly analogous to the test of the slope. Suppose, for example, we have a one-sided situation with null and alternative hypotheses

$$H_0 : \mu = \mu_0 \quad ,$$

$$H_1 : \mu > \mu_0 \quad .$$

We use the test criterion $T = (\hat{\mu}-\mu_0)/S(\hat{\mu})$, and reject H_0 at the significance level α if we observe $t \geq t_{n-1}(1-\alpha)$. In actual fact this test is equivalent to the corresponding test of β_1, as in Section 9.1 since $\mu = \beta_1 x$ and $\hat{\mu} = B_1 x$.

9.3. Test of goodness of fit of the model

When we have replicate observations on Y for each of several x values, we may use these to test whether or not the variances are inversely proportional to the assumed weights if these are at all in doubt. If we accept the weighting scheme, we may proceed to test whether or not a particular type of polynomial function for μ, $\mu = f(x)$, is valid.

Suppose that our data consist of r sets, the ith set at x_i comprising the n_i observations $\{Y_{ij}\}$, $j = 1,2,...n_i$, such that $Y_{ij} \sim N(\mu_i,\sigma_i^2)$ or $Y_{ij} \sim N(\mu_i,\sigma_{0,i}^2/w_i)$, where $\sigma_{0,i}^2 = w_i\sigma_i^2$. According to our model, this quantity is the same for all sets, say σ^2. and $\mu_i = \beta_1 x_i$, so that, in examining the goodness of fit, we seek to test whether or not our data are consistent with the assumption that $Y_{ij} \sim N(\beta_1 x_i,\sigma^2/w_i)$.

Firstly, we examine the variances under the hypotheses:

$$H_0 : \sigma_{0,1}^2 = \sigma_{0,2}^2 = \dots \sigma_{0,r}^2 = \sigma^2, \text{ say,}$$

$$H_1 : \text{not } H_0 \ .$$

The null hypothesis states that all the σ_0^2's are equal whereas the alternative is a quite general hypothesis allowing that some, but not necessarily all, are unequal. Clearly, the sum of squares about each set mean $\bar{Y}_i = \sum_j Y_{ij}/n_i$ provides an estimator of the variance of that set, σ_i^2 , thus

$$s_i^2 = \sum_j (Y_{ij} - \bar{Y}_i)^2/(n_i - 1) . \tag{1}$$

Hence

$$s_{O,i}^2 = \sum_j w_i (Y_{ij} - \bar{Y}_i)^2/(n_i - 1) \tag{2}$$

provides an estimator of $\sigma_{O,i}^2$, and it is thus a simple matter to use Bartlett's test (for example) to check the homogeneity of the $\{\sigma_{O,i}^2\}$. If the data pass the test, we may pool the $\{s_{O,i}^2\}$ to obtain an estimator of σ^2 from all the sets, thus

$$s_a^2 = \Sigma(n_i - 1)s_{O,i}^2/\Sigma(n_i - 1)$$

$$= \sum_i\sum_j w_i (Y_{ij} - \bar{Y}_i)^2/(n - r) , \tag{3}$$

where $n = \sum_i n_i$. Further transformation gives

$$s_a^2 = \{\sum_i\sum_j w_i Y_{ij}^2 - \sum_i w_i (\sum_j Y_{ij})^2/n_i\}/(n - r) , \tag{4}$$

which is more convenient for calculation. If the data do not pass the test, then the basis of the weighting scheme needs reconsideration. It is important to emphasize that we cannot proceed to the second stage of our examination of the goodness of fit without a soundly based set of weights.

In the second stage of our test, we obtain an independent estimator of σ^2 from the mean square of the residuals about the regression, and compare this with s_a^2 obtained previously from the pooled variances of the replicates about their means. Our two hypotheses are

$$H_O : \mu = \beta_1 x ,$$

$$H_1 : \text{not } H_O .$$

We should mention that our H_1 here essentially calls for a more complicated model than that of H_O since the case where

$\mu = 0$ is included in H_0 by putting $\beta_1 = 0$. We shall see in Chapter 13 how to test more complicated models against H_0.

If H_0 is true, we may estimate β_1 using the triads $\{w(\bar{Y}_i), x_i, \bar{Y}_i\}$, where $w(\bar{Y}_i)$ denotes the weight associated with $\bar{Y}_i$. We put

$$w(\bar{Y}_i) = \frac{\sigma^2}{\sigma^2(\bar{Y}_i)} = \frac{\sigma^2}{\sigma_i^2/n_i}$$

$$= n_i w_i \quad ,$$

and so

$$B_1 = \frac{\Sigma n_i w_i x_i \bar{Y}_i}{\Sigma n_i w_i x_i^2} \quad .$$

Our estimator of σ^2 from the scatter of our data points about the regression is

$$S_b^2 = \Sigma n_i w_i (\bar{Y}_i - B_1 x_i)^2/(r - 1) \quad . \tag{5}$$

More conveniently, for computational purposes, we have

$$S_b^2 = \{\sum_i w_i (\sum_j Y_{ij})^2/n_i - (\sum_i\sum_j w_i x_i Y_{ij})^2/\Sigma n_i w_i x_i^2\}/(r - 1) . \tag{6}$$

From equations (4) and (6), we see that the sums of squares that are divided by the appropriate numbers of degrees of freedom to give S_a^2 and S_b^2 are differences involving the three sums of squares

$$\sum_i\sum_j w_i Y_{ij}^2 \quad , \qquad \sum_i w_i (\sum_j Y_{ij})^2/n_i \quad ,$$

and

$$(\sum_i\sum_j w_i x_i Y_{ij})^2/\Sigma n_i w_i x_i^2 \quad .$$

This last quantity is also the sum of squares due to the regression of $\{Y_{ij}\}$ on $\{x_i\}$. Thus, for the present purposes, we do not need to calculate B_1 and the residuals, but simply go directly to these three sums of squares and derive S_a^2 and S_b^2 from them.

Clearly, S_a^2 is an unbiased estimator of σ^2 whatever form the function $\mu = f(x)$ takes. However, although S_b^2 is always independent of S_a^2, it is only an unbiased estimator of σ^2 under H_0; under H_1, its expectation will be larger than σ^2. Hence, the test ratio

$$T = S_b^2/S_a^2$$

is distributed as $F_{r-1,n-r}$ under H_0, but tends to be larger when H_1 applies. Thus, we use a one-sided test, rejecting H_0 at a significance level α if $t \geq F_{r-1,n-r}(1-\alpha)$.

One very important point in regard to replication must be made. It is essential that all the sources of variation contributing to the mean square residual about the regression are present when replicate experiments are made. For example, suppose there is a day-to-day variation which influences the observations (perhaps temperature, light intensity, etc.) and that replicate experiments are carried out in a single day followed by study of the x dependence of Y on several succeeding days. Any significant difference found between the S_a^2 and S_b^2 will reflect the variation in the diurnal factors in addition to any deviations of a fundamental kind from the model. In such a case, a better procedure would be to obtain one replicate for each x on each day.

9.4. Comparison of two or more sets of data

If two or more sets of data are available, it is often of considerable importance to test whether or not each set provides an estimator, B_1, of the same parameter, β_1. Two rather different situations are shown diagrammatically in Figure 12.2 for the case where 2 sets A and B are available. If the test supports the hypothesis that each set provides an estimate of the same quantity β_1, we describe the experimental slopes as homogeneous. In this case, we shall want to know the simplest procedure for calculating revised estimators of β_1 and $\sigma(B_1)$ from all the data. We shall describe a procedure which is suitable for any number of sets of data. The special case of two sets of data can be treated more simply as outlined in Section 9.4.4.

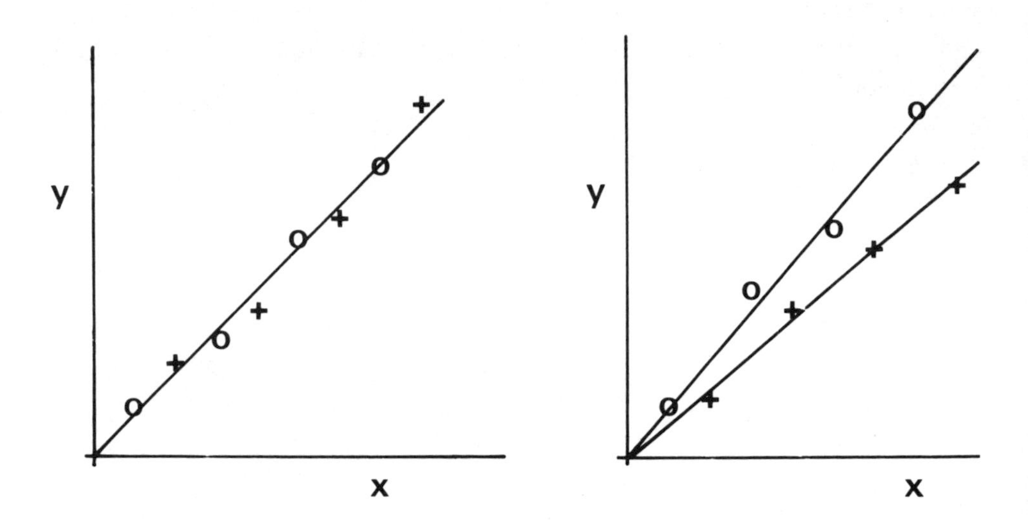

Figure 12.2. Illustration of (i) homogeneous and (ii) heterogeneous sets of data, A and B; o A + B

We shall use (w_{ij}, x_{ij}, Y_{ij}) to represent the experimental data for the jth point of the ith set of data; σ_{ij}^2 stands for the variance of Y_{ij}; w_{ij} is the weight associated with Y_{ij} and

$$\sigma_i^2 = w_{ij}\sigma_{ij}^2 \quad , \tag{1}$$

independent of j, i = 1 to r, j = 1 to n_i.

We shall take it that we have available for each set the following sums

$$\left.\begin{aligned} S_{xY,i} &= \sum_j w_{ij}x_{ij}Y_{ij} \\ S_{xx,i} &= \sum_j w_{ij}x_{ij}^2 \\ S_{YY,i} &= \sum_j w_{ij}Y_{ij}^2 \end{aligned}\right\} \tag{2}$$

From these, we may easily calculate the estimator of β_1 from the ith set, namely B_{1i}, and the corresponding weighted sum of squared residuals, $\sum_j w_{ij}D_{ij}^2$. The quantity B_{1i} is given by the usual expression

$$B_{1i} = \sum_j w_{ij} x_{ij} Y_{ij} / \sum_j w_{ij} x_{ij}^2$$

$$= S_{xY,i}/S_{xx,i} \ . \qquad (3)$$

The weighted sum of squared residuals, which we shall represent as $S_{DD,i}$, is

$$S_{DD,i} = \sum_j w_{ij} D_{ij}^2 \ ,$$

where

$$D_{ij} = Y_{ij} - B_{1i} x_{ij} \ .$$

In terms of the sums defined in equation (2), $S_{DD,i}$ may be written

$$S_{DD,i} = S_{YY,i} - S_{xY,i}^2/S_{xx,i} \ . \qquad (4)$$

9.4.1. Stage 1: testing the equality of the σ_i^2's

For each set, we have a quantity σ_i^2, given by equation (9.4.1). The data of each set enable an estimator of this quantity to be calculated, thus

$$s_i^2 = S_{DD,i}/(n_i - 1) \ .$$

Now these estimators are independent and so we may use Bartlett's test (see Chapter 9) to decide between the two alternative hypotheses

$$H_0 : \sigma_1^2 = \sigma_2^2 = \ldots = \sigma_r^2 = \sigma^2 \ ,$$

$$H_1 : \text{not } H_0 \ .$$

Alternatively, we may simply test the ratio of the largest to the smallest of the s_i^2's by an F_{max} test as in Chapter 9, Section 3.1.

9.4.2. Stage 2: testing that the experimental slopes 'agree'

Having tested and accepted, or assumed, that our σ_i^2's are consistent (otherwise we cannot proceed), we now specify our two hypotheses concerning the slopes. These are

$$H_0 : \beta_{11} = \beta_{12} = \dots \beta_{1r} = \beta_1 ,$$

$$H_1 : \text{not } H_0 .$$

We now explore the implications of H_0. If this hypothesis is correct, we have a set of estimators, $\{B_{1i}\}$, all with the same expectation β_1, though of different variances $\{\sigma^2(B_{1i})\}$. In fact

$$\sigma^2(B_{1i}) = \sigma_i^2/\sum_j w_{ij}x_{ij}^2 ,$$

$$= \sigma^2/\sum_j w_{ij}x_{ij}^2 ,$$

$$= \sigma^2/S_{xx,i} .$$

We take as our estimator of β_1, the weighted mean, $\bar{B}_1$, of the individual estimators,

$$\bar{B}_1 = \Sigma w(B_{1i})B_{1i}/\Sigma w(B_{1i}) , \qquad (1)$$

where $w(B_{1i}) = \sigma^2/\sigma^2(B_{1i})$ is the weight associated with the ith estimator, B_{1i}. Hence our estimator is given by

$$\bar{B}_1 = \Sigma S_{xx,i}B_{1i}/\Sigma S_{xx,i} .$$

Now if the parent Y_{ij}'s are normally distributed, the B_{1i}'s are distributed as $N\{\beta_1,\sigma^2(B_{1i})\}$. It can be shown that

$$\Sigma(B_{1i} - \bar{B}_1)^2/\sigma^2(B_{1i})$$

is distributed as χ^2 with r-1 degrees of freedom. Hence, an estimator of σ^2 is provided by the quantity, S_a^2, where

$$S_a^2 = \Sigma S_{xx,i}(B_{1i} - \bar{B}_1)^2/(r - 1) . \tag{2}$$

This estimator of σ^2 is calculated from the scatter of the experimental slopes about the mean slope.

An independent estimator of σ^2 may be calculated from the scatter of the experimental points about the fitted lines. If the Y_{ij}'s are normally distributed, we have seen that

$$\sum_j w_{ij}D_{ij}^2/\sigma_i^2 = \sum_j w_{ij}D_{ij}^2/\sigma^2$$

is distributed as χ^2 with n_i-1 degrees of freedom. Hence, using the additivity principle,

$$\sum_i\sum_j w_{ij}D_{ij}^2/\sigma^2$$

is distributed as χ^2 with $\Sigma(n_i-1)$ degrees of freedom, because the r components of the sum over i are independently distributed. Hence a second estimator of σ^2 may be calculated from our data, thus

$$S_b^2 = \sum_i\sum_j w_{ij}D_{ij}^2/\Sigma(n_i - 1) ,$$

$$= \Sigma S_{DD,i}/(n - r) , \tag{3}$$

where $n = \Sigma n_i$.

Thus, under H_O, the test ratio

$$T = S_a^2/S_b^2 \tag{4}$$

is distributed as $F_{r-1,n-r}$ and hence the acceptability of H_O may be decided by the usual single-tail F-test. We should note that under the alternative hypothesis H_1, $E(S_a^2) > \sigma^2$ because the true individual slopes of each set are not coincident.

If we accept H_O, we conclude that the individual set slopes are merely estimates of the same quantity and hence we may proceed to Stage 3 in which we seek to combine the data of all sets in the most useful and economical way.

9.4.3. Stage 3: computation of pooled estimators from all the data

If the data pass the test for homogeneity, it only remains to calculate estimators of β_1 and σ^2. Our estimator of β_1 is the weighted mean of the individual slopes, equation (9.4.2.1.)

$$\bar{B}_1 = \Sigma S_{xx,i} B_{1i} / \Sigma S_{xx,i} .$$

This can be recast using the definitions of $S_{xx,i}$ and B_{1i} given in equations (9.4.1) and (9.4.2). The result is

$$\bar{B}_1 = \sum_i \sum_j w_{ij} x_{ij} Y_{ij} / \sum_i \sum_j w_{ij} x_{ij}^2 .$$

$\bar{B}_1$ is thus the value which would be calculated if no regard were paid to the fact that the data belong to different sets. This is just what we should expect if the data were homogeneous.

Three different estimators of σ^2 may be derived from the data. These are

$$s_a^2 = \Sigma S_{xx,i} (B_{1i} - \bar{B}_1)^2 / (r - 1) , \qquad (1)$$

$$s_b^2 = \Sigma S_{DD,i} / (n - r) , \qquad (2)$$

$$s_c^2 = \sum_i \sum_j w_{ij} (Y_{ij} - \bar{B}_1 x_{ij})^2 / (n - 1) . \qquad (3)$$

Equations (1) and (2) are simply repeats of equations (9.4.2.2) and (9.4.2.3) while equation (3) is the application of equation (6.1.3) to the particular case here with the residuals calculated from the mean slope. Equation (3) is the obvious choice when the data are homogeneous, and it may be more conveniently derived from the following

$$s_c^2 = \sum_i \sum_j w_{ij} Y_{ij}^2 - \Big(\sum_i \sum_j w_{ij} x_{ij} Y_{ij}\Big)^2 / \sum_i \sum_j w_{ij} x_{ij}^2$$

$$= \Sigma S_{YY,i} - (\Sigma S_{xY,i})^2 / \Sigma S_{xx,i} . \qquad (4)$$

The variance of the estimator, $\bar{B}_1$, is given by the usual expression

$$\sigma^2(\bar{B}_1) = \sigma^2/\Sigma w(B_{1i}) ,$$

$$= \sigma^2/\Sigma S_{xx,i} .$$

Hence, an estimator of $\sigma^2(\bar{B}_1)$ is

$$S^2(\bar{B}_1) = S^2/\Sigma S_{xx,i} ,$$

where the expression for S_c^2 may be substituted for S^2. This estimator is associated with (n-1) degrees of freedom and may be used to calculate confidence intervals on β_1 in the usual way.

If the data do not pass the test, the $\{B_{1i}\}$ afford estimators of the separate slopes. Either way, S_b^2 is an unbiased estimator of σ^2 for all sets.

9.4.4. The special case where only two sets of data are to be compared

In this case, Stage 1 of the previous sequence may be simplified since we have only two estimators of σ^2 which we may compare using a two-sided F-test. Stage 2 may also be simplified. The two hypotheses to be considered reduce to

$$H_O : \beta_{11} = \beta_{12} = \beta_1, \text{ say},$$

$$H_1 : \beta_{11} \neq \beta_{12} ,$$

and hence, under H_O ,

$$E(B_{11} - B_{12}) = 0 .$$

We also recognize that

$$S(B_{11} - B_{12}) = \{S^2(B_{11}) + S^2(B_{12})\}^{\frac{1}{2}}$$

$$= S(1/\Sigma w_{1j}x_{1j}^2 + 1/\Sigma w_{2j}x_{2j}^2)^{\frac{1}{2}} .$$

The estimator, S, of σ may be formulated in terms of the residuals D_{1j} and D_{2j}. Hence the test ratio

$$T = \frac{B_{11} - B_{12}}{S(B_{11} - B_{12})}$$

$$= \frac{(B_{11} - B_{12})(n_1 + n_2 - 2)^{\frac{1}{2}}}{(\Sigma w_{1j}D_{1j}^2 + \Sigma w_{2j}D_{2j}^2)^{\frac{1}{2}}(1/\Sigma w_{1j}x_{1j}^2 + 1/\Sigma w_{2j}x_{2j}^2)^{\frac{1}{2}}}$$

$$= \frac{(B_{11} - B_{12})(n - 2)^{\frac{1}{2}}}{\{\Sigma S_{YY,i} - \Sigma(S_{xY,i}^2/S_{xx,i})\}^{\frac{1}{2}}\{\Sigma(1/S_{xx,i})\}^{\frac{1}{2}}}, \quad i = 1,2 .$$

The random variable T will be distributed as t_{n-2} under H_0. Thus, the acceptability of H_0 can be decided. The appropriate estimators of β_1 and σ^2 follow as before.

Example 1. Equilibrium between a solution of an involatile solute and solvent vapour at some fixed partial pressure is established at a higher temperature than in the case of a pure solvent. All theories predict that, when the solution is sufficiently dilute, the difference in temperature, ΔT, is proportional to the concentration, c, of the solute, that is,

$$\Delta T = kc, \text{ at low } c, \qquad (1)$$

where k is a constant. In this example, we seek to establish that this relationship holds for solutions of diphenyl in benzene up to a concentration of 9.118 kg m^{-3}.

In the apparatus used here, the temperature difference is proportional to a potential difference, V, which is the measured quantity. To test the goodness of fit of equation (1), it is sufficient to test that the data fit the model

$$Y = \beta_1 x + \varepsilon , \qquad (2)$$

where Y and x stand for suitably coded transforms of V and c, β_1 is a constant proportional to k, and ε is the error term, which we assume to be distributed normally with zero expectation and constant variance.

The procedure of Section 9.3 was followed. Seven solutions of diphenyl in benzene were made, ranging in concentration from 2.894 to 9.118 kg m^{-3}. Samples were taken from these solutions in random order over a period of several days and placed in the apparatus. Three replicate determinations of potential difference were made in succession for each solution. The means of the three observations for the various solutions on the various days are presented in Table 12.1. We shall refer to these means as the 'raw' data.

Table 12.1. The 'raw' data after taking out the units

Index Number of solution	$\frac{m}{g}$	$x = \frac{c}{kg\ m^{-3}}$	$Y = V/\mu Volt$ Day 1	Day 2	Day 3	Day 4
1	0.1447	2.894	$316.\dot{6}$	315.0	$311.\dot{6}$	315.0
2	0.2010	4.020	$453.\dot{3}$	435.0	$438.\dot{3}$	435.0
3	0.2580	5.160	$563.\dot{3}$	$566.\dot{6}$	555.0	555.0
4	0.3076	6.152	$678.\dot{3}$	$688.\dot{3}$	670.0	$683.\dot{3}$
5	0.3571	7.142	$778.\dot{3}$	$786.\dot{6}$	$766.\dot{6}$	$778.\dot{3}$
6	0.4049	8.098	885.0	$878.\dot{3}$	$873.\dot{3}$	$873.\dot{3}$
7	0.4559	9.118	985.0	$991.\dot{6}$	$983.\dot{3}$	$986.\dot{6}$

m is the mass of diphenyl weighed out and dissolved in 50 cm^3 of solution

---

Suppose we represent a data point by (x_i, y_{ij}), where $i = 1,2,\ldots 7$ (the solution number) and $j = 1,2,3,4$ (the day number). Our first task is to test for the homogeneity of the variance of $\{Y_{ij}\}$. We calculate the means $\bar{y}_i = \sum_j y_{ij}/4$ and the estimates of variance $s_i^2 = \sum_j (y_{ij} - \bar{y}_i)^2/3$, for $i = 1,2,\ldots 7$. These are shown in Table 12.2.

Table 12.2. Sample means and variances for individual solutions based on 3 degrees of freedom

i	$\bar{y}_i$	s_i^2
1	314.6	4.40
2	440.4	76.6
3	560.0	35.2
4	680.0	61.1
5	777.5	67.6
6	877.5	30.6
7	986.7	13.0

The number of significant figures entered here is in accordance with our usual rule.

A simple F_{max} test using the ordinary F tables to show the homogeneity of the σ_i^2 is inconclusive in this case and so we may either use the proper tables or Bartlett's test. The latter test shows that at the 5 percent significance level we accept the hypothesis that the σ_i^2's are the same, say σ^2. Hence, we may pool the individual estimates to obtain our first estimate of σ^2, say

$$s_a^2 = 41.2$$

with 21 degrees of freedom.

Our second estimate of σ^2 comes from the scatter of the $\bar{y}_i$ about the regression. We could go directly to equation (9.3.6), but it will probably be more instructive to first calculate b_1, the least squares estimate of the slope of the line through all the data points. We have

$$b_1 = \Sigma n_i x_i \bar{y}_i / \Sigma n_i x_i^2$$

$$= \Sigma x_i \bar{y}_i / \Sigma x_i^2 ,$$

since n_i is the same for all i. We find

$$b_1 = \frac{31409.4809}{288.732} = 108.78 .$$

Other statistical information concerning this particular straight line is collected together in Table 12.3. We observe from this table that the mean square residual about this line is 27.7, that is

$$\hat{\sigma}^2(\bar{Y}) = 27.7 .$$

Since each $\bar{Y}$ is the mean of 4 points, we obtain as our second estimate of σ^2, the error variance of the Y's,

$$\hat{\sigma}^2(Y) = s_b^2 = 4 \times 27.7 = 110.8 ,$$

with 6 degrees of freedom. Our test criterion is

$$t = \frac{110.8}{41.2}$$

$$= 2.69 ,$$

which is just greater than $F_{6,21}(0.95) = 2.57$. Hence, we reject at the 5 percent significance level the hypothesis that our data fit the simple linear model, $\mu = \beta_1 x$.

This result is surprising in view of what we know about the theory of solutions and the way in which the apparatus behaves. To seek to explain this anomaly, we examine the fit of the data through the column of residuals $\bar{y}_i - \hat{\mu}_i$. We observe that the residual for the fourth solution is considerably larger than for the others indicating that this experimental point lies relatively well off the fitted line; in fact, this point contributes nearly 70 percent of the total sum of squares about the regression. Thus this particular point is largely responsible for the decision (a marginal one anyway) to reject H_0. In actual fact, if we omit the data for this solution from our analysis, we obtain a test criterion $t = 26.4/37.9 = 0.70 < F_{5,18}(0.95)$, showing that the majority of our data

Table 12.3 Statistical information about the mean line through the data points (truncated in most cases to five significant figures)

S_w	$= 7$	(6)
S_x	$= 4.2584 \times 10^1$	(3.6432×10^1)
$S_{\bar{y}}$	$= 4.6367 \times 10^3$	(3.9567×10^3)
S_{xx}	$= 2.8873 \times 10^2$	(2.5088×10^2)
$S_{\bar{y}\bar{y}}$	$= 3.4170 \times 10^6$	(2.9546×10^6)
$S_{x\bar{y}}$	$= 3.1409 \times 10^4$	(2.7226×10^4)
S_{dd}	$= 1.6620 \times 10^2$	(3.2954×10^1)
b_1	$= 1.0878 \times 10^2$	(1.0852×10^2)
$s(B_1)$	$= 3.0974 \times 10^{-1}$	(1.6208×10^{-1})
s^2	$= 27.70$	(6.59)
ν	$= 6$	(5)
r	$= 9.9980 \times 10^{-1}$	(9.9997×10^{-1})

Sum of squares explained by regression = 3.4168×10^6 (2.9546×10^6)

i	x_i	$\bar{y}_i$	$\hat{\mu}_i$	$\bar{y}_i - \hat{\mu}_i$	$(\hat{\mu}_i)$	$(\bar{y}_i - \hat{\mu}_i)$
1	2.894	314.6	314.82	-0.22	(314.06)	(0.54)
2	4.020	440.4	437.31	3.09	(436.25)	(4.15)
3	5.160	560.0	561.33	-1.33	(559.96)	(0.04)
4	6.152	680.0	669.24	10.76	-	-
5	7.142	777.5	776.94	0.56	(775.05)	(2.45)
6	8.098	877.5	880.93	-3.43	(878.80)	(-1.30)
7	9.118	986.7	991.89	-5.19	(989.49)	(-2.79)

is consistent with the model. (The statistical analysis for the 6-point mean line is shown in brackets in Table 12.3). The evidence points to a source of variation between solutions which is not included in the factors causing variability in the work within solutions.

If this is the case, our experimental design is invalid since to test our model it is essential to make sure that the same factors causing variability are present in both the replication experiments and the investigation of the concentration dependence of the $\{\bar{Y}_i\}$. If the same factors are not present, our so-called replicates are pseudo-replicates and do not serve to estimate the variance of the distribution from which the observations on the concentration dependence were drawn. This exemplifies the comment we made in Chapter 1 when discussing experimental design, where we said that it is all too easy to draw observations from a narrower, less representative distribution than we require.

Obviously the next step in our investigation would be to reassess our procedure to see whether or not all the sources of variation have been included in our replication work. Although it is out of place here to discuss the details, we should note that at least one source of variation present in the concentration studies does not contribute to the variability between replicates. Each solution was made by weighing the appropriate mass of solute, dissolving it in the solvent, and then making up to a nominal 50 cm^3 in volume in a graduated volumetric flask. Any variation in volume between the seven graduated flasks used in this work would not influence the replication experiments within solutions but would certainly contribute to the variability about the regression; in particular, a relatively large difference in volume between the flask used for solution four and the remaining flasks would give rise to the effect discussed earlier. This is something that would be considered further in a full investigation.

Example 2. A typical situation in which the equations of Section 4.3 may be applied is the construction and use of a calibration line. Here, we consider data obtained in the course of an investigation of the degree of conversion of high alumina cement, a problem which we mentioned briefly in Chapter 11. Samples of cement with given degrees of conversion, f, were dehydrated by heating and a quantity Y relating to the energy absorbed in this process was measured. According to a simple, crude theory, Y should be directly proportional to the quantity 100/f - 1, in the absence of experimental error. Hence, we may write our observations in the usual form

$$Y = \beta_1 x + \varepsilon , \tag{1}$$

where Y is the observed quantity, x is 100/f - 1, β_1 is a constant to be estimated, and ε is the random error. Table 12.4 shows the data and Figure 12.3 shows the least squares line with $b_1 = 0.670$, $s(B_1) = 0.044$, $S^2 = 6.74 \times 10^{-3}$, the estimates of variance being based on 3 degrees of freedom.

Table 12.4. Calibration data for samples of high alumina cement of known degrees of conversion

f	x	y
40	1.50	0.932
50	1.00	0.742
67	0.493	0.397
95	0.053	0.108

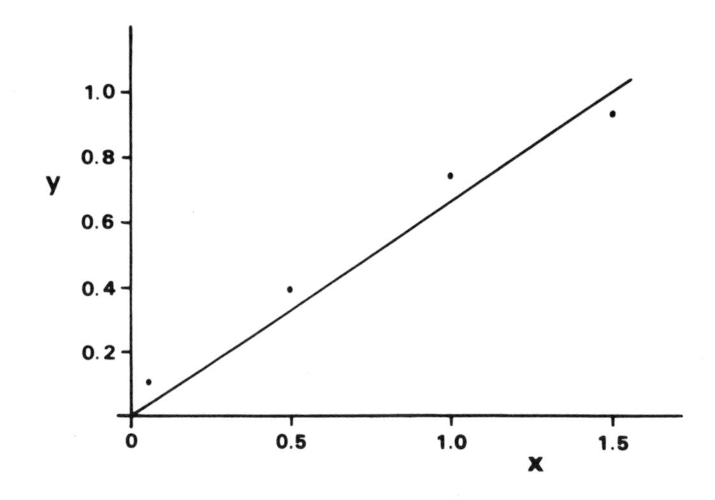

Figure 12.3 Least squares line for the data of Table 12.4.

The figure shows that equation (1) is a fairly poor approximation to the data so that elaboration of the model to take account of possible differences in variance between the Y_i is unwarranted. However, equation (1) does provide a basis for estimating x and $\sigma(x)$ for samples of unknown conversion. Suppose, for example, we obtain y = 0.454 for such a sample. Then, by equations (4.3.2) and (4.3.3)

$$\hat{x} = y/b_1 = 0.454/0.670 = 0.678$$

and

$$s(\hat{x}) = s(1 + \hat{x}^2/\Sigma x_i^2)^{\frac{1}{2}}/b_1 = 0.129 .$$

Our usual rule indicates that we take $\hat{x}$ = 0.68 with standard error 0.13. To obtain the corresponding quantities for $\hat{f}$ and $s(\hat{f})$, we use

$$\hat{f} = 100/(\hat{x} + 1) = 100/1.68 = 59.52$$

and

$$s(\hat{f}) = 10^{-2}\hat{f}^2 s(\hat{x}) = 4.61 .$$

Thus, the estimated degree of conversion from our measured y, 0.454, is 59.5 with standard error 4.6 based on three degrees of freedom.

We remark that we have deliberately chosen a difficult physico-chemical system with which to illustrate inverse interpolation since this must be typical of many situations. Our example shows that, notwithstanding the indifferent model used for the data, it is possible to obtain some idea of the true location of the interpolated value.

CHAPTER 13

THE POLYNOMIAL THROUGH THE ORIGIN OR THROUGH SOME OTHER FIXED POINT

A natural extension of the discussion of Chapter 12 is the problem of fitting a polynomial in x through a set of data points $\{x_i, y_i\}$, the polynomial being constrained to pass either through the origin or through some other fixed point. As we explained previously, the latter case is easily dealt with by shifting the origin of the co-ordinate system.

The procedures described in this chapter assume that the Y's are independent and that E(Y) can be expressed as a polynomial in x. However, we may not always know the appropriate order of polynomial. Consequently, we shall have to discuss not only the topics raised in the previous chapter, but also the criterion by which the most appropriate order of the polynomial may be selected.

1. Formal statement of the model

We shall assume that the random variable, Y, is related to x by the equation

$$Y = \sum_{p=1}^{m} \beta_p x^p + \varepsilon , \qquad (1)$$

where m is the degree of polynomial, the set $\{\beta_p\}$ are constants, and ε is the error term. We shall assume that

$$E(\varepsilon) = 0 \quad \text{for all } x \qquad (2)$$

and

$$V(\varepsilon) = \sigma_i^2 \text{ for } x = x_i . \qquad (3)$$

We shall not discuss the special case where the variance of the error term is constant at each of the data points, since this is included in the more general situation. It follows from equations (1), (2) and (3) that

$$E(Y) = \mu = \Sigma\beta_p x^p , \qquad (4)$$

$$V(Y_i) = \sigma_i^2 \quad \text{at } x = x_i \ . \tag{5}$$

As in Chapter 12, it is convenient to introduce a set of weights, $\{w_i\}$, and a constant, σ^2, which is such that

$$w_i = \frac{\sigma^2}{\sigma_i^2} \ . \tag{6}$$

2. Estimation using the method of least squares

The least squares method enables us to find estimators of the unknown parameters, $\{\beta_p\}$, μ, and σ^2. We choose the estimators $\{B_p\}$ of $\{\beta_p\}$ so that the weighted sum of squares, \$, given by

$$\$ = \sum_i w_i (Y_i - \sum_p \beta_p x_i^p)^2 \ , \tag{1}$$

is a minimum when $\beta_p = B_p$, for each β. We thus solve the m simultaneous equations

$$\frac{\partial \$}{\partial \beta_p} = 0 \ , \quad p = 1 \text{ to } m \ , \tag{2}$$

i.e.
$$\sum_i w_i x_i^p (Y_i - \sum_q B_q x_i^q) = 0 \ , \qquad p = 1 \text{ to } m \ , \tag{3}$$

where we have changed the index of the inner summation to avoid confusion. These are called the normal equations and are sufficient to determine uniquely the set of m unknowns, $\{B_p\}$, and hence $\hat{\mu}$ from

$$\hat{\mu} = \Sigma B_p x^p \ . \tag{4}$$

It is apparent that two distinct types of summation over i are involved. First, there are terms of the type

$$\Sigma w_i x_i^p Y_i$$

and second, there are terms involving x_i and w_i alone, namely

$$\Sigma w_i x_i^p x_i^q \ .$$

The essential simplicity of the set of normal equations is more obvious if we represent these sums by the symbols θ_p and ϕ_{pq} as follows

$$\theta_p = \Sigma w_i x_i^p Y_i \ , \tag{5}$$

$$\phi_{pq} = \Sigma w_i x_i^p x_i^q \ . \tag{6}$$

In terms of these symbols, the normal equations appear thus

$$\sum_q B_q \phi_{pq} = \theta_p \ , \quad p = 1 \text{ to } m. \tag{7}$$

It is very important to appreciate that all the quantities θ_p and ϕ_{pq} are known explicitly for a given set of data $\{w_i, x_i, y_i\}$. Thus the set of equations represented by (7) is simply a set of <u>m simultaneous</u>, <u>linear</u> equations in the m unknowns, $\{B_p\}$. Hence these equations can be solved by the classical methods for such problems. One such method involves the successive elimination of the unknowns until only one remains; the remainder can then be obtained by reversing the process. Another method makes use of the properties of matrices. Formally, the normal equations may be expressed as

$$\underline{\Phi}\underline{B} = \underline{\Theta} \ , \tag{8}$$

where $\underline{\Phi}$ is a symmetrical m x m matrix and $\underline{B}$ and $\underline{\Theta}$ are column vectors of order m x 1. In this notation, the vector of solutions is

$$\underline{B} = \underline{\Phi}^{-1}\underline{\Theta} \ , \tag{9}$$

where $\underline{\Phi}^{-1}$ represents the inverse of the matrix $\underline{\Phi}$. We shall not consider these methods in detail since there are many excellent texts dealing with these procedures. Furthermore, we expect that many of our readers will have access to a library of computer programs which will contain sub-routines for solving just such problems.

Although the estimation problem is now solved, further

development to derive $\{V(B_p)\}$, $V(\hat{\mu})$, etc. tends to become rather cumbersome for reasons which will be clearer later. To simplify these later derivations, we reformulate our problem in terms of orthogonal polynomials in x.

3. Orthogonal polynomials in x

In this section, we digress from our main theme to introduce a new idea, the idea that a set of numbers such as $\{x_i\}$ can be transformed into another set of numbers, $\{T_p(x_i)\}$ where p is an integer, which have the property

$$\sum_i w_i T_p(x_i) T_q(x_i) = 0 , \tag{1}$$

when $p \neq q$. For the purposes of this chapter, these numbers are generated by substituting $x = x_i$ in expressions of the type

$$T_p(x) = c_{p1}x + c_{p2}x^2 + \ldots x^p , \tag{2}$$

where c_{pq}, the coefficient of x^q in the expansion of $T_p(x)$, is a constant to be determined. We shall require m of these polynomials, namely

$$\left.\begin{aligned} T_1(x) &= x \\ T_2(x) &= c_{21}x + x^2 \\ T_3(x) &= c_{31}x + c_{32}x^2 + x^3 \\ &\vdots \\ T_p(x) &= c_{p1}x + c_{p2}x^2 + c_{p3}x^3 + \ldots x^p \\ &\vdots \\ T_m(x) &= c_{m1}x + c_{m2}x^2 + c_{m3}x^3 + \ldots x^m . \end{aligned}\right\} \tag{3}$$

We shall refer to this set of polynomials (not to be confused with the fitted polynomials discussed in the previous sections) as the set of orthogonal polynomials, although <u>the orthogonality condition is only fulfilled at the data points</u>. It should be noted that we have defined the coefficient of the highest power of x in a given polynomial to be unity, i.e. $c_{pp} = 1$, $p = 1$ to m. The more compact form of equation (3) is

$$\underline{T} = \underline{C}\underline{X} , \tag{4}$$

where $\underline{C}$ is the lower triangular matrix of coefficients, c_{pq}, thus

$$\underline{C} = \begin{bmatrix} 1 & 0 & 0 & \cdots & 0 \\ c_{21} & 1 & 0 & \cdots & 0 \\ c_{31} & c_{32} & 1 & \cdots & 0 \\ \vdots & \vdots & \vdots & & \vdots \\ c_{m1} & c_{m2} & c_{m3} & \cdots & 1 \end{bmatrix} . \tag{5}$$

$\underline{X}$ and $\underline{T}$ stand for the column vectors $(x, x^2, x^3, \ldots x^m)'$ and $(T_1, T_2, T_3, \ldots T_m)'$, where T_1, T_2, etc. represent $T_1(x)$, $T_2(x)$ etc., and a prime denotes transpose.

It can be seen from equation (5) that there are $m(m-1)/2$ coefficients, c_{pq}, to be determined. Since the pairing of the subscripts, p and q, in the orthogonality condition

$$\sum_i w_i T_p(x_i) T_q(x_i) = 0, \; p \neq q \; ,$$

is restricted to precisely the same combinations as in c_{pq}, it follows that the number of orthogonality conditions is exactly equal to the number of coefficients to be determined. In Appendix 2, we show and exemplify how to determine the coefficients and orthogonal polynomials.

4. Reformulation of the model

We now return to the main theme of this chapter, that is the process of fitting a polynomial through a set of points subject to the constraint that it must pass through some fixed point, e.g. the origin. To overcome the difficulties to which we referred briefly in Section 2, we reformulate our model for the random variable, Y, in terms of a subset of $T_p(x)$, thus

$$Y = \sum_{p=1}^{m} \alpha_p T_p(x) + \varepsilon . \tag{1}$$

As before, m stands for the degree of polynomial and ε for the error term; the $\{\alpha_p\}$, p = 1, 2,... m, are constants and the $\{T_p(x)\}$ are defined by equations (3.1) and (3.2). Of course, the random variable ε has just the same properties as defined by equations (1.2) and (1.3), since equation (1) is merely an alternative way of writing equation (1.1). We see that

$$E(Y) = \mu = \Sigma\alpha_p T_p(x) \tag{2}$$

and

$$V(Y_i) = \sigma_i^2 \text{ at } x = x_i . \tag{3}$$

It is useful to consider the expectation of our random variable Y at each of the data points. For the ith point,

$$\mu_i = \sum_q \alpha_q T_q(x_i) . \tag{4}$$

Note we have changed the index from p to q for reasons which will become clear shortly. Suppose that we multiply both sides of equation (4) by $w_i T_p(x_i)$ and sum over i, giving

$$\begin{aligned} \sum_i w_i \mu_i T_p(x_i) &= \sum_{iq}\sum \alpha_q w_i T_p(x_i) T_q(x_i) \\ &= \sum_q \alpha_q \sum_i w_i T_p(x_i) T_q(x_i) . \end{aligned}$$

The term on the right-hand side vanishes when $p \neq q$, because of the orthogonality property of the $\{T_q(x_i)\}$. Hence,

$$\sum_i w_i \mu_i T_p(x_i) = \alpha_p \sum_i w_i T_p^2(x_i) , \tag{5}$$

or

$$\alpha_p = \frac{\sum_i w_i \mu_i T_p(x_i)}{\sum_i w_i T_p^2(x_i)} . \tag{6}$$

It is important to appreciate that equations (5) and (6) can never serve to give the numerical values of the set $\{\alpha_p\}$, since the numerical values of the μ_i will remain unknown. The purpose of the derivation is to derive an expression [equation (5)] which we shall need subsequently.

5. Estimation using the method of least squares

The least squares method enables us to find estimators of the unknowns which appear in our reformulated model, viz. $\{\alpha_p\}$, μ and σ^2. We choose the estimators $\{A_p\}$ of $\{\alpha_p\}$ so that the weighted sum of squares, \$, given by

$$\$ = \sum_i w_i \{Y_i - \sum_p \alpha_p T_p(x_i)\}^2 \tag{1}$$

is a minimum when $\alpha_p = A_p$, for each p. As before, we solve the m simultaneous equations

$$\frac{\partial \$}{\partial \alpha_p} = 0, \quad p = 1 \text{ to } m\ , \tag{2}$$

that is,

$$\sum_i w_i T_p(x_i)\{Y_i - \sum_q A_q T_q(x_i)\} = 0, \quad p = 1 \text{ to } m\ . \tag{3}$$

It is at this point that the trouble which we took in transforming the $\{x_i\}$ into the orthogonal polynomials $\{T_q(x_i)\}$ really starts to pay off. On multiplying out equation (3), all terms containing $T_p(x_i)T_q(x_i)$, $p \neq q$, disappear and we obtain for the least squares estimator of the pth coefficient of equation (4.1)

$$A_p = \frac{\sum_i w_i Y_i T_p(x_i)}{\sum_i w_i T_p^2(x_i)}\ . \tag{4}$$

We note in passing that, in the special case of m = 1 (i.e. the straight line passing through the origin) where $T_1(x_i) = x_i$, we recover equation (6.1.3) of Chapter 12, viz.

$$A_1 = B_1 = \Sigma w_i x_i Y_i / \Sigma w_i x_i^2\ .$$

5.1. Expectations of the estimators, $\{A_p\}$, and $\hat{\mu}$

Equation (4) of the previous section shows that each of the estimators $\{A_p\}$ is a linear combination of the $\{Y_i\}$. Hence

$$E(A_p) = \frac{\sum_i w_i \mu_i T_p(x_i)}{\sum_i w_i T_p^2(x_i)}$$

$$= \alpha_p \text{ , by equation (4.6) ,}$$

showing that $\{A_p\}$ are unbiased.

Since

$$\hat{\mu} = \Sigma A_p T_p(x) \text{ ,}$$

it follows immediately that

$$E(\hat{\mu}) = \Sigma \alpha_p T_p(x) = \mu \text{ ,}$$

showing that $\hat{\mu}$ is an unbiased estimator of μ.

5.2. The covariance matrix of the $\{A_p\}$ and the variance of $\hat{\mu}$

The (p,q) element of the covariance matrix of $\{A_p\}$ is

$$E\{(A_p-\alpha_p)(A_q-\alpha_q)\} \text{ .}$$

We substitute for A_p and A_q from equation (5.4) and for α_p and α_q from equation (4.6). The (p,q) element of our matrix thus becomes

$$E\left|\frac{\left\{\sum_i w_i(Y_i-\mu_i)T_p(x_i)\right\}\left\{\sum_j w_j(Y_j-\mu_j)T_q(x_j)\right\}}{\left\{\sum_i w_i T_p^2(x_i)\right\}\left\{\sum_j w_j T_q^2(x_j)\right\}}\right| \text{ .}$$

Here the denominator is a constant containing no random variable. Consider the numerator, K, say

$$K = E\left\{\sum_i\sum_j w_i w_j (Y_i-\mu_i)(Y_j-\mu_j)T_p(x_i)T_q(x_j)\right\} \text{ .}$$

When $i \neq j$, $E\{(Y_i-\mu_i)(Y_j-\mu_j)\} = 0$, since Y_i and Y_j are independent, and, when $i = j$,

$$E\{(Y_i-\mu_i)^2\} = \sigma_i^2 = \sigma^2/w_i \; .$$

Hence K becomes

$$K = \sigma^2 \sum_i w_i T_p(x_i) T_q(x_i) \; ,$$

so that

$$C(A_p,A_q) = \frac{\sigma^2 \sum_i w_i T_p(x_i) T_q(x_i)}{\left\{\sum_i w_i T_p^2(x_i)\right\}\left\{\sum_i w_i T_q^2(x_i)\right\}} \; . \tag{1}$$

When $p \neq q$, this vanishes because of the orthogonality condition. Hence

$$C(A_p,A_q) = 0 \tag{2}$$

when $p \neq q$, and

$$V(A_p) = \frac{\sigma^2 \sum_i w_i T_p^2(x_i)}{\left\{\sum_i w_i T_p^2(x_i)\right\}^2} = \frac{\sigma^2}{\sum_i w_i T_p^2(x_i)} \; . \tag{3}$$

Since

$$\hat{\mu} = \Sigma A_p T_p(x) \; ,$$

$$V(\hat{\mu}) = \Sigma T_p^2(x) V(A_p) \; ,$$

because the covariances are zero. Thus

$$V(\hat{\mu}) = \sigma^2 \sum_p \frac{T_p^2(x\;)}{\sum_i w_i T_p^2(x_i)} \; , \tag{4}$$

using equations (2) and (3).

5.3. Comments on $E(A_p)$ and $V(A_p)$

It is very important to remember that the random variables $\{Y_i\}$ are defined on a series of sampling spaces fixed by the particular values of x_i. Equally, the polynomials $\{T_q(x_i)\}$ are fixed by the same particular values of x_i. Consequently the estimators, $\{A_p\}$, are associated with a definite subset of the general sampling space $S = \{x,y\}$. Thus we cannot expect the expectations and variances of corresponding elements of two sets $\{A_p\}$ and $\{A_p'\}$ associated with different sets $\{x_i\}$ and $\{x_i'\}$ to be the same.

5.4. The weighted sum of squared residuals, and derived variance estimators

We commence by deriving an expression for the weighted sum of squared residuals

$$\Sigma w_i D_i^2 = \Sigma w_i (Y_i - \hat{\mu}_i)^2$$

$$= \Sigma w_i Y_i^2 - 2\Sigma w_i Y_i \hat{\mu}_i + \Sigma w_i \hat{\mu}_i^2 \quad . \qquad (1)$$

Now the quantity $\Sigma w_i Y_i \hat{\mu}_i$ is equal to $\Sigma w_i \hat{\mu}_i^2$. To show this, we first note that equation (5.3) yields

$$\sum_i w_i T_p(x_i) Y_i = \sum_i w_i T_p(x_i) \hat{\mu}_i \ .$$

We then multiply each side of this equation by A_p and sum over p to obtain

$$\sum_p \sum_i w_i A_p T_p(x_i) Y_i = \sum_p \sum_i w_i A_p T_p(x_i) \hat{\mu}_i \ ,$$

giving the result

$$\Sigma w_i Y_i \hat{\mu}_i = \Sigma w_i \hat{\mu}_i^2 \ .$$

Hence equation (1) becomes

$$\Sigma w_i D_i^2 = \Sigma w_i Y_i^2 - \Sigma w_i \hat{\mu}_i^2 \ .$$

We note that

$$\Sigma w_i \hat{\mu}_i^2 = \sum_i w_i \sum_{pq}\sum A_p A_q T_p(x_i) T_q(x_i)$$

$$= \sum_{pq}\sum A_p A_q \sum_i w_i T_p(x_i) T_q(x_i)$$

$$= \sum_i w_i \sum_p A_p^2 T_p^2(x_i) ,$$

since the terms involving $p \neq q$ vanish by the orthogonality condition. Hence

$$\left.\begin{aligned} \Sigma w_i D_i^2 &= \Sigma w_i Y_i^2 - \sum_i w_i \sum_p A_p^2 T_p^2(x_i) \\ &= \Sigma w_i Y_i^2 - \sum_p A_p^2 \sum_i w_i T_p^2(x_i) . \end{aligned}\right\} \qquad (3)$$

Thus, $\Sigma w_i \hat{\mu}_i^2$ or $\sum_p A_p^2 \sum_i w_i T_p^2(x_i)$ is the sum of squares due to the regression, and $\Sigma w_i D_i^2$, the sum of squares about the regression, is the difference between the total sum of squares and the sum of squares due to the regression.

Taking expectations of both sides of equation (2) and recalling that

$$V(Y_i) = E(Y_i^2) - \mu_i^2 ,$$

$$V(\hat{\mu}_i) = E(\hat{\mu}_i^2) - \mu_i^2 ,$$

leads to the result

$$E(\Sigma w_i D_i^2) = \Sigma w_i V(Y_i) - \Sigma w_i V(\hat{\mu}_i) \qquad (4)$$

$$= \sum_i \sigma^2 - \sum_i w_i \sigma^2 \sum_q \frac{T_q^2(x_i)}{\sum_i w_i T_q^2(x_i)} ,$$

using $V(Y_i) = \sigma^2/w_i$, and equation (5.3.3). Hence

$$E(\Sigma w_i D_i^2) = \sum_i \sigma^2 - \sum_q \sigma^2 \qquad (5)$$

$$= (n-m)\sigma^2 . \qquad (6)$$

We draw attention to the remarkable simplicity of equation (5). The origin of its simplicity is easily identified by retracing the arguments. We see that each random variable Y_i contributes σ^2 to the expectation of $\Sigma w_i D_i^2$, and that each estimator required in the estimation of μ reduces the expectation of $\Sigma w_i D_i^2$ by σ^2. Hence if there are n random variables Y_i and m estimators A_q or B_q, the expectation of $\Sigma w_i D_i^2$ is $(n-m)\sigma^2$.

It follows from equation (6) that an unbiased estimator of σ^2 is S^2, with n-m degrees of freedom, given by

$$S^2 = \Sigma w_i D_i^2/(n-m) \; . \tag{7}$$

Having now estimated σ^2 by S^2, we may conveniently estimate the variances of $\{A_p\}$ and $\hat{\mu}$ by

$$S^2(A_p) = \frac{S^2}{\sum_i w_i T_p^2(x_i)} \; , \tag{8}$$

$$S^2(\hat{\mu}) = S^2 \sum_p \frac{T_p^2(x)}{\Sigma_i w_i T_p^2(x_i)} \; . \tag{9}$$

We note that as $x \to 0$, $T_p(x) \to 0$ (all p) and so $S(\hat{\mu})$, the error of interpolation, tends to zero in conformity with the constraint imposed by our model that the polynomial must pass through the origin.

5.5. Altering the order of the polynomial

In many instances, the degree of the polynomial to be fitted to the data will not be known in advance. Very often the reason an experimentalist wishes to fit a polynomial to his data is no stronger than the desire to 'account' for a departure from linearity. So having tried, say, a quadratic fit, his natural curiosity prompts him to enquire what would happen if he were to choose a cubic. Later we shall present a method of testing the goodness of fit of different-order polynomials when a normal distribution may be assumed for the

Y's (see Section 10.2). For the moment, however, we are concerned with a more limited problem, namely how to estimate the coefficients of the new model:

$$\left.\begin{aligned} Y &= \sum_{p=1}^{m+1} \alpha'_p T_p(x) + \varepsilon' , \\ E(\varepsilon') &= 0 , \\ V(\varepsilon') &= \sigma'^2_i \text{ for } x = x_i , \end{aligned}\right\} \tag{1}$$

in which the highest power of x is now m+1, and the primes serve to distinguish quantities appropriate to this model from those of our former model, viz.

$$\left.\begin{aligned} Y &= \sum_{p=1}^{m} \alpha_p T_p(x) + \varepsilon , \\ E(\varepsilon) &= 0 , \\ V(\varepsilon) &= \sigma^2_i \text{ for } x = x_i . \end{aligned}\right\} \tag{2}$$

As we point out in Appendix 2, the inclusion of a higher-order orthogonal polynomial in the set $\{T_p(x_i)\}$ in no way affects the original elements, and therefore we have not primed the T_p's. Let us now look at the estimators of μ, α_p , and α'_p. These are given by the expressions

new model

$$\hat{\mu}' = \sum_{p=1}^{m+1} A'_p T_p(x) , \tag{3}$$

$$A'_p = \frac{\sum_i w_i Y_i T_p(x_i)}{\sum_i w_i T^2_p(x_i)} , \quad p = 1 \text{ to } m+1 , \tag{4}$$

old model

$$\hat{\mu} = \sum_{p=1}^{m} A_p T_p(x) , \tag{5}$$

$$A_p = \frac{\sum_i w_i Y_i T_p(x_i)}{\sum_i w_i T^2_p(x_i)} , \quad p = 1 \text{ to } m . \tag{6}$$

Inspection of equations (4) and (6) shows that

$$A'_p = A_p , \quad p = 1 \text{ to } m . \tag{7}$$

This means in practice that it is not necessary to recalculate the first m estimators for an (m+l)-order polynomial if they have already been calculated for a polynomial of order m; only the extra estimator A'_{m+1} needs to be calculated. Furthermore, the primes are clearly redundant since the values of these estimators are independent of the order of the fitted polynomial. We thus write the estimator of μ in our new model in the form

$$\hat{\mu}' = \sum_{p=1}^{m+1} A_p T_p(x) . \tag{8}$$

Comparison of this equation with equation (5) shows that

$$\hat{\mu}' = \hat{\mu} + A_{m+1} T_{m+1}(x) . \tag{9}$$

In other words, we may obtain the estimator of μ in our new model [the (m+l)-order polynomial] by simply adding the term $A_{m+1}T_{m+1}(x)$ to the estimator of μ which we calculated in our original model (the polynomial of order m). In short, altering the order of the fitted polynomial requires nothing more than the computation of a new set of polynomials, $\{T_{m+1}(x_i)\}$, and A_{m+1}.

It should be appreciated that the simplicity of this procedure is entirely a consequence of the formulation of the problem in terms of the orthogonal polynomials in x. In contrast, if we used the original approach through the power series representation

$$\hat{\mu} = \Sigma B_p x^p ,$$

when the order of the polynomial is changed, all the coefficients $\{B_p\}$ have to be recalculated.

5.5.1. The weighted sum of squared residuals and derived variance estimators

The weighted sum of squared residuals for the new model is easily calculated. We have from equation (5.4.3)

$$\Sigma w_i D_i'^2 = \Sigma w_i Y_i^2 - \sum_{p=1}^{m+1} A_p^2 \sum_i w_i T_p^2(x_i) \qquad (1)$$

$$= \Sigma w_i D_i^2 - A_{m+1}^2 \sum_i w_i T_{m+1}^2(x_i) \ . \qquad (2)$$

Hence, we do not have to recalculate $\Sigma w_i D_i'^2$ from 'scratch'; we need only work out $A_{m+1}^2 \sum_i w_i T_{m+1}^2(x_i)$ and subtract this from our former sum of squared residuals for the polynomial of order m to find the new value appropriate to the order m+1.

As before, we estimate the basic error variance, σ'^2, appropriate to the new model using

$$S'^2 = \frac{\Sigma w_i D_i'^2}{n-(m+1)} \ .$$

From this, we obtain new estimators of $V(A_p)$ and $V(\hat{\mu})$ by replacing S^2 by S'^2 in equations (5.4.8) and (5.4.9).

6. Return to the original polynomial $\hat{\mu} = \Sigma B_p x^p$

Although we prefer to work in terms of the orthogonal polynomials $\{T_p(x_i)\}$ and the $\{A_p\}$, it is often necessary to represent the fitted polynomial in terms of the $\{B_p\}$. If we have used the orthogonal polynomial approach, we can best achieve this objective by deriving a relationship between the $\{A_p\}$ and the $\{B_p\}$. An extension of this derivation will lead us to the elements of the covariance matrix, $C(B_p, B_q)$.

We commence by writing down the two alternative expressions which we have used for E(Y). These are

$$\mu = \sum_p \beta_p x^p \ , \qquad (1)$$

$$\mu = \sum_p \alpha_p T_p(x) \ . \qquad (2)$$

If we substitute for $T_p(x)$ in terms of the coefficients c_{pq} and x^q, we obtain

$$\mu = \sum_{p=1}^{m} \alpha_p (\sum_{q=1}^{p} c_{pq} x^q) . \tag{3}$$

Now compare the coefficients of x^r in equations (1) and (3). We obtain

$$\beta_r = \sum_{p=1}^{m} \alpha_p c_{pr} , \qquad r = 1 \text{ to } m.$$

But when $r > p$, all the coefficients $c_{pr} = 0$ (see Appendix 2), since there are no higher terms in $T_p(x)$ than x^p. Hence

$$\beta_r = \sum_{p=r}^{m} \alpha_p c_{pr} .$$

Finally, we replace the dummy suffix r by q to obtain

$$\beta_q = \sum_{p=q}^{m} \alpha_p c_{pq} . \tag{4}$$

It follows directly (or alternatively by comparing the coefficients of x^r in the alternative formulations of $\hat{\mu}$) that

$$B_q = \sum_{p=q}^{m} A_p c_{pq} . \tag{5}$$

Equation (5) enables us to compute the estimators $\{B_q\}$ from the estimators $\{A_p\}$ and the coefficients $\{c_{pq}\}$.

6.1. Means and covariance matrix of $\{B_q\}$

Obviously

$$E(B_q) = \sum_{p=q}^{m} E(A_p) c_{pq}$$

$$= \sum_{p=q}^{m} \alpha_p c_{pq}$$

Hence from equation (6.4) we obtain

$$E(B_q) = \beta_q \, . \tag{1}$$

This shows that B_q is an unbiased estimator of β_q.

To obtain $C(B_p,B_q)$ our reasoning is similar to that used to derive $C(A_p,A_q)$. We have

$$C(B_p,B_q) = E\left[\left\{\sum_{r=p}^{m} c_{rp}(A_r-\alpha_r)\right\}\left\{\sum_{s=q}^{m} c_{sq}(A_s-\alpha_s)\right\}\right]$$

$$= \sum_{r=p}^{m}\sum_{s=q}^{m} c_{rp}c_{sq}C(A_r,A_s) \, .$$

Since $C(A_r,A_s) = 0$ for $r \neq s$, this becomes

$$C(B_p,B_q) = \sum_{r=\max(p,q)}^{m} c_{rp}c_{rq}V(A_r)$$

$$= \sigma^2 \sum_{r=\max(p,q)}^{m} \{c_{rp}c_{rq}/\sum_i w_i T_r^2(x_i)\} \, , \tag{2}$$

which may be estimated by

$$s^2 \sum_{r=\max(p,q)}^{m} \{c_{rp}c_{rq}/\sum_i w_i T_r^2(x_i)\} \, . \tag{3}$$

We note that $C(B_p,B_q) \neq 0$ for $p \neq q$. For $p = q$ we have

$$V(B_q) = \sigma^2 \sum_{r=q}^{m} \{c_{rq}^2/\sum_i w_i T_r^2(x_i)\}, \tag{4}$$

which may be estimated, again, by replacing σ^2 by s^2.

6.2. Why bother with the power series representation?

It may be asked why we should trouble to return to the power series representation when the problem of fitting the data has already been solved. There are a number of reasons. First, it will often happen that the coefficients $\{\beta_p\}$ have some physical significance and hence computation of estimates of them is of considerable importance; in such cases, it will also be valuable to compute the estimates of variance $\{s^2(B_p)\}$.

A second reason is equally important. The power series representation enables us to see quickly whether the estimated polynomial curve takes a shape which is in conflict with accepted theory, either within or outside the range of the data points. The first two differential coefficients with respect to x and a graph of the curve are often useful tools for this purpose, in particular for consideration of maxima, minima, points of inflexion, etc. When a high speed computer is available it is all too easy to fit a meaningless equation to a set of data points, so care is needed.

A third reason for calculating the coefficients of the power series is implied by Section 5.3. When we wish to examine whether or not two or more sets of data based on <u>different</u> sets of x values have the same means μ, expressed as polynomials in x, it will usually be more convenient to think in terms of testing the homogeneity of the $\{\beta_p\}$ sets using the $\{B_p\}$ sets to do so.

7. A modification of the model

Our present model takes the form

$$\left.\begin{aligned} Y &= \Sigma\beta_p x^p + \varepsilon , \\ E(\varepsilon) &= 0 , \\ V(\varepsilon) &= \sigma_i^2 \text{ at } x = x_i . \end{aligned}\right\} \qquad (1)$$

So far we have been concerned with establishing the properties of the estimators $\{B_p\}$ (and the associated quantities $\{A_p\}$). It may happen that not all the $\{\beta_p\}$ are unknowns. A possible situation is that β_1 is known as a result of a theoretical analysis of the way in which the true value, μ, depends on x.

We commence by formally defining the new model. We write

$$\left.\begin{aligned} Y &= kx + \sum_{p=2}^{m} \beta_p x^p + \varepsilon , \\ E(\varepsilon) &= 0 , \\ V(\varepsilon) &= \sigma_i^2 \text{ at } x = x_i , \end{aligned}\right\} \quad (2)$$

where k, a known constant, replaces β_1. We define a new random variable, Z, thus

$$Z = Y/x - k . \quad (3)$$

We see that

$$E(Z) = \mu/x - k , \quad (4)$$

$$V(Z) = V(Y)/x^2 . \quad (5)$$

Equations (3), (4), and (5) taken together with equation (2) lead to the following conclusions about the new random variable, Z ,

$$Z = \sum_{p=1}^{m'} \beta'_p x^p + \varepsilon' , \quad (6)$$

$$E(\varepsilon') = 0 , \quad (7)$$

$$V(\varepsilon') = \sigma_i'^2 \text{ at } x = x_i . \quad (8)$$

Each of the primed quantities is related to the corresponding unprimed quantities in a simple way:

(i) $m' = m - 1$,

(ii) $\beta'_p = \beta_{p+1}$, $p = 1$ to $m-1$,

(iii) $\varepsilon' = \varepsilon/x$,

(iv) $\sigma_i'^2 = \sigma_i^2/x_i^2$.

Equations (6), (7), (8) are identical to those of the original model, equations (1.1), (1.2), (1.3), except for the presence of the primes. Hence, we may use the previous arguments to estimate the parameters $\{\beta'_p\}$. Of course, a certain amount of initial transformation is necessary. We have to compute a new set of weighting factors to go with $\{Z_i\}$ from the original weighting factors which we had associated with $\{Y_i\}$. Since $\sigma_i'^2 = \sigma_i^2/x_i^2$,

$$w'_i = w_i x_i^2 . \tag{9}$$

Thus if all the weights associated with $\{Y_i\}$ were unity, the new weights associated with $\{Z_i\}$ would be $\{x_i^2\}$.

8. Sampling distributions assuming the normality of the data

As we saw in Chapter 12, no further progress can be made without assuming something about the distribution of the error term and therefore of $\{Y_i\}$. We assume

$$\varepsilon \sim N(0,\sigma_i^2) \text{ at } x = x_i ,$$

which is equivalent to assuming that Y_i is normally distributed about μ_i with variance σ_i^2, $i = 1$ to n.

8.1. The distributions of $\{A_p\}$, $\{B_p\}$, $\hat{\mu}$ and $\Sigma w_i D_i^2/\sigma^2$

We recall equation (5.4) for the pth member of the set $\{A_p\}$.

$$A_p = \sum_i w_i Y_i T_p(x_i) / \sum_i w_i T_p^2(x_i) .$$

Now as we mentioned in Chapter 4, any linear combination of normally distributed random variables is itself normally distributed. Consequently, the distribution of the estimator A_p is normal; we may thus write

$$A_p \sim N\left\{\alpha_p, \sigma^2 / \sum_i w_i T_p^2(x_i)\right\}$$

from the results of Sections 5.1 and 5.2. Precisely similar arguments lead to the conclusions that

$$B_p \sim N\left[\beta_p, \sigma^2 \sum_{q=p}^{m} \{c_{qp}^2/\sum_i w_i T_q^2(x_i)\}\right]$$

and

$$\hat{\mu} \sim N\left[\mu, \sigma^2 \sum_p \{T_p^2(x)/\sum_i w_i T_p^2(x_i)\}\right] .$$

It can be shown (see Chapter 16, Section 2) that $\Sigma w_i D_i^2/\sigma^2$ is distributed as χ_{n-m}^2 independently of the $\{A_p\}$.

9. Calculation of confidence intervals for $\{\alpha_p\}$, $\{\beta_p\}$ and μ

Since $(A_p - \alpha_p)/\sigma(A_p)$ and $\Sigma w_i D_i^2/\sigma^2$ are independently distributed as $N(0,1)$ and χ_{n-m}^2 respectively, it follows that

$$\frac{A_p - \alpha_p}{S(A_p)} \sim t_{n-m} .$$

Hence a confidence interval for α_p with coefficient $1-\alpha$ is given by

$$a_p \pm t_{n-m}(1-\alpha/2)s(A_p) , \tag{1}$$

where a_p and $s(A_p)$ are realizations of A_p and $S(A_p)$ for particular data.

More usually we shall require confidence intervals for β_p and μ. Similar arguments show that

$$\frac{B_p - \beta_p}{S(B_p)} \quad \text{and} \quad \frac{\hat{\mu} - \mu}{S(\hat{\mu})}$$

are both distributed as t_{n-m} and yield confidence intervals for β_p and μ of

$$b_p \pm t_{n-m}(1-\alpha/2)s(B_p) , \tag{2}$$

$$\hat{\mu} \pm t_{n-m}(1-\alpha/2)s(\hat{\mu}) . \tag{3}$$

For the particular values $\{\mu_i\}$, we may alternatively use the interval

$$y_i \pm t_{n-m}(1-\alpha/2)s/w_i^{\frac{1}{2}} , \tag{4}$$

since $(Y_i-\mu_i)/(S/w_i^{\frac{1}{2}}) \sim t_{n-m}$ also. However, we would ordinarily expect this to be wider than interval (3), except for larger x_i.

10. Tests of a number of hypotheses

We now seek to answer a similar series of questions to those of Chapter 12, Section 9, by performing tests on corresponding hypotheses concerning the unknown parameters. We are unlikely to want to test hypotheses relating to $\{\alpha_p\}$, since such a set corresponds only to a particular set of x_i and so is of rather restricted relevance. However, it will be clear from the following sections how to perform tests on $\{\alpha_p\}$ should these be required.

10.1. Tests on $\{\beta_p\}$ and μ

Here, we consider tests relating to the set of coefficients $\{\beta_p\}$ and μ. We illustrate the procedure for β_q for which we assume that we have some theoretical value available denoted by $\beta_{q,0}$. Our hypotheses are

$$H_0 : \beta_q = \beta_{q,0} \qquad \left| \begin{array}{l} H_1 : \beta_q > \beta_{q,0} , \\ H_1 : \beta_q < \beta_{q,0} , \quad \text{or} \\ H_1 : \beta_q \neq \beta_{q,0} . \end{array} \right.$$

Under H_0, the test criterion

$$T = \frac{B_q - \beta_{q,0}}{S(B_q)}$$

is distributed as t_{n-m}. Hence, depending on whether the test is one- or two-sided, we compare the realization of T or |T|

with the appropriate critical value of t_{n-m} as explained in Chapter 8. When the value of $\beta_{q,0}$ is zero, this test is equivalent to enquiring whether or not the term x^q makes a significant contribution to μ.

For tests on μ, we proceed analogously using

$$T = (\hat{\mu}-\mu_0)/S(\hat{\mu})$$

as our test criterion, μ_0 being the null hypothetical value.

10.2. Test of goodness of fit of the model

The procedure employed here is precisely the same as that described in Chapter 12. We obtain n_i replicate observations $\{Y_{ij}\}$, $j = 1,2,...n_i$ at x_i, $i = 1,2,...r$, and from these obtain suitable estimators of σ^2. If the model is correct, $Y_{ij} \sim N(\sum_p \beta_p x_i^p, \sigma^2/w_i)$, where the $\{\beta_p\}$ and $\sigma^2 = w_i\sigma_i^2$ are the same for all sets. There are thus two distinct stages in checking the goodness of fit of our data to the model. These are first, verification that the variances of the r sets of $\{Y_{ij}\}$ may be taken as inversely proportional to the assumed weights, and second, checking that the distribution means of the $\{Y_{ij}\}$, that is $\{\mu_i\}$, are correctly given by $\sum_p \beta_p x_i^p$.

The first stage is exactly the same as that described in Chapter 12. Assuming that we accept the weighting scheme, we obtain our first estimator of σ^2, with n-r degrees of freedom, as

$$S_a^2 = \sum_{ij}\sum w_i(Y_{ij} - \bar{Y}_i)^2/(n - r)$$

$$= \{\sum_{ij}\sum w_i Y_{ij}^2 - \Sigma w_i n_i \bar{Y}_i^2\}/(n - r)$$

$$= \{\sum_{ij}\sum w_i Y_{ij}^2 - \Sigma w(\bar{Y}_i)\bar{Y}_i^2\}/(n - r) ,$$

where $w(\bar{Y}_i) = w_i n_i$ is the weight associated with $\bar{Y}_i$. Now if we represent these sums by suitably suffixed S's (appropriate weights being understood), we obtain simply

$$s_a^2 = (S_{YY} - S_{\bar{Y}\bar{Y}})/(n - r) .$$

We now proceed to the next stage of our test. Our two hypotheses are

$$H_0 : \mu = \sum_{p=1}^{m} \beta_p x^p ,$$

$$H_1 : \text{not } H_0 .$$

As in Chapter 12, a lower order polynomial is included in H_0 (by putting some β's equal to zero) and is not appropriate to H_1. Under H_0, we can obtain a second estimator of σ^2 from the mean square residual about the regression $\hat{\mu} = \sum_{p=1}^{m} B_p x^p$ using the r triads $\{w(\bar{Y}_i), x_i, \bar{Y}_i\}$. Then the sum of squared residuals is given by

$$\Sigma w(\bar{Y}_i) D_i^2 = S_{\bar{Y}\bar{Y}} - S_{AATT} ,$$

where

$$S_{AATT} = \sum_p A_p^2 \sum_i w(\bar{Y}_i) T_p^2(x_i) .$$

Thus, our second estimator of σ^2 with r-m degrees of freedom is

$$s_b^2 = (S_{\bar{Y}\bar{Y}} - S_{AATT})/(r - m) .$$

Accordingly we perform an F test using the criterion

$$T = s_b^2/s_a^2 ,$$

which will be distributed as $F_{r-m,n-r}$ under H_0. Since $E(s_b^2) > \sigma^2$ under H_1, while $E(s_a^2) = \sigma^2$ under both hypotheses, a one-sided test is appropriate. Hence we reject H_0 at the α significance level if our realization t exceeds the $1-\alpha$ quantile, $F_{r-m,n-r}(1-\alpha)$.

10.3. Choice of the order of the polynomial

So far, we have been fitting to our data a polynomial of order m, a given integer. Sometimes, we may have theoretical grounds to decide what m should be. More often, we shall not have such prior knowledge so we need to use the data to decide what order of polynomial we should use. This section is concerned with possible procedures.

10.3.1. Use of replicates

The previous section provides us with one method. If we have a number of replicates available, such that n-r, the number of degrees of freedom of S_a^2, is adequate (say at least five or six), then we test the goodness of fit of a polynomial of order 1 to the data; if the fit is significantly poor (that is, if we reject H_0), we test the fit of a polynomial of order 2; if, again, the result is significant, we pass to order 3, and so on until we reach an order for which we get a non-significant result. This is the order we choose.

10.3.2. Comparison of polynomials

Often, however, we shall have either no replicates at all, or too few replicates, for the previous method to be applied. We shall now describe a procedure whereby we may again test successively increasing orders of polynomials until we reach a non-significant result. However, this time we are not testing the goodness of fit of each polynomial to the data. Instead, we compare two consecutive order polynomials to see whether the inclusion of an extra power in x effects a significant reduction in the sum of squares about the regression relative to the estimated σ^2 for the higher order polynomial. If there is a significant reduction, we proceed to a further stage, testing the higher order of our two original polynomials against a polynomial one order higher still, and so on until we reach a non-significant result.

Suppose we are comparing two polynomials of order m and m+1. From the results of Section 5.5.1, the sums of squared residuals for the two polynomials are

$$S_{DD,m} = (\Sigma w_i D_i^2)_m = \Sigma w_i Y_i^2 - \sum_{p=1}^{m} A_p^2 \sum_i w_i T_p^2(x_i)$$

and

$$S_{DD,m+1} = (\Sigma w_i D_i^2)_{m+1} = \Sigma w_i Y_i^2 - \sum_{p=1}^{m+1} A_p^2 \sum_i w_i T_p^2(x_i) .$$

The <u>extra</u> sum of squares, R_{m+1}, accounted for by the higher order polynomial is thus,

$$R_{m+1} = S_{DD,m} - S_{DD,m+1} = A_{m+1}^2 \; \Sigma w_i T_{m+1}^2(x_i) .$$

Now the two hypotheses which we wish to test are

$$H_0 : \mu = \sum_{p=1}^{m} \alpha_p T_p(x) ,$$

$$H_1 : \mu = \sum_{p=1}^{m+1} \alpha_p T_p(x) .$$

Under H_0, it can be shown that $S_{DD,m+1}/\sigma^2$ and R_{m+1}/σ^2 are independently distributed as χ^2_{n-m-1} and χ^2_1 respectively. Hence, under H_0, the test criterion

$$T = \frac{R_{m+1}}{S_{DD,m+1}/(n-m-1)}$$

will be distributed as $F_{1,n-m-1}$. Under H_1, it is still true that $S_{DD,m+1}$ and R_{m+1} are independent and that $S_{DD,m+1}/\sigma^2$ is distributed as χ^2_{n-m-1}. However, R_{m+1}/σ^2 will tend to be greater than one if H_1 applies and so a one-sided test on T is appropriate here. Thus, we reject H_0 at the significance level α if $t \geq F_{1,n-m-1}(1-\alpha)$, the $1-\alpha$ quantile of $F_{1,n-m-1}$.

The reader may find it helpful to prove the results given in Table 13.1 which shows the expectations of the various sums of squares used in the above analysis. For this purpose, the relation

$$E(X^2) = V(X) + \{E(X)\}^2 ,$$

which holds for any random variable, X, will be found useful. It should also be understood that H_0 implies $\alpha_{m+1} = 0$.

Table 13.1. Expectations of sums of squares under the null and alternative hypotheses

Sum of squares ss	E(ss) Under H_0	Under H_1
$S_{DD,m}$	$(n-m)\sigma^2$	$(n-m)\sigma^2 + \alpha_{m+1}^2 \sum_i w_i T_{m+1}^2(x_i)$
$S_{DD,m+1}$	$(n-m-1)\sigma^2$	$(n-m-1)\sigma^2$
R_{m+1}	σ^2	$\sigma^2 + \alpha_{m+1}^2 \sum_i w_i T_{m+1}^2(x_i)$

To consolidate the arguments, we give in detail the first two stages of the above procedure. First, we test between the hypotheses

$$H_0 : \mu = 0 ,$$

$$H_1 : \mu = \alpha_1 T_1(x) ,$$

using

$$S_{DD,0} = \Sigma w_i Y_i^2 ,$$

$$S_{DD,1} = \Sigma w_i Y_i^2 - A_1^2 \Sigma w_i T_1^2(x_i) ,$$

so that

$$R_1 = A_1^2 \Sigma w_i T_1^2(x_i) .$$

Our test criterion is

$$T = R_1/\{S_{DD,1}/(n - 1)\}.$$

If the result is significant, we test between the hypotheses

$$H_0 : \mu = \alpha_1 T_1(x) ,$$

$$H_1 : \mu = \alpha_1 T_1(x) + \alpha_2 T_2(x) ,$$

using the test criterion

$$T = R_2/\{S_{DD,2}/(n - 2)\}.$$

We continue this process until we obtain our first non-significant result. Then if R_r was our last significant numerator, we take r as the order of our polynomial.

We should mention that this procedure may lead to a non-significant result being obtained 'too-early', and a final order for the polynomial being chosen which is lower than the true value. Suppose that we are comparing two polynomials of order m and m+1, but that the true order is m+2. Following the arguments which lead to the conclusions summarised in Table 13.1, we may show that if this were the case

$$E(S_{DD,m}) = (n-m)\sigma^2 + \alpha_{m+1}^2 \Sigma w_i T_{m+1}^2(x_i) + \alpha_{m+2}^2 \Sigma w_i T_{m+2}^2(x_i) ,$$

$$E(S_{DD,m+1}) = (n-m-1)\sigma^2 + \alpha_{m+2}^2 \Sigma w_i T_{m+2}^2(x_i) ,$$

and

$$E(R_{m+1}) = \sigma^2 + \alpha_{m+1}^2 \Sigma w_i T_{m+1}^2(x_i) .$$

Thus, the test ratio

$$T = \frac{R_{m+1}}{S_{DD,m+1}/(n-m-1)}$$

may yield a non-significant result due to the denominator overestimating σ^2, leading us to accept the order m, 2 powers of x less than is correct. For this reason, some authors suggest that one should not settle upon a final order until at least two consecutive non-significant results have been obtained, then choosing as the final order that which gave the last significant numerator. There is clearly some sense in this procedure.

10.3.3. What happens if too high an order is chosen?

We may wonder what would be the effect on our estimator $\hat{\mu}$, if we should accidentally accept a polynomial of higher order than the true order by one of these procedures. Suppose the true order were m_1, and that we had concluded that it was $m_2 > m_1$. Then

$$\hat{\mu} = \sum_{p=1}^{m_2} A_p T_p(x) \; ,$$

while

$$\mu = \sum_{p=1}^{m_1} \alpha_p T_p(x) \; .$$

Now

$$E(\hat{\mu}) = \sum_{p=1}^{m_2} E(A_p) T_p(x)$$

$$= \sum_{p=1}^{m_1} \alpha_p T_p(x) \; , \text{ since } \alpha_p = 0, \; p > m_1 \; ,$$

$$= \mu \; .$$

Also

$$V(\hat{\mu}) = \sigma^2 \sum_{p=1}^{m_2} \frac{T_p^2(x)}{\sum_i w_i T_p^2(x_i)} > \sigma^2 \sum_{p=1}^{m_1} \frac{T_p^2(x)}{\sum_i w_i T_p^2(x_i)} \; .$$

Hence $\hat{\mu}$ remains an unbiased estimator, but its variance is larger than it would have been if the correct order had been chosen.

10.4. Comparison of two or more sets of data

It may happen that we have a number of different sets of data through which we have fitted polynomials of degree m. Naturally, the m estimators, $\{B_p\}$, p = 1 to m, will vary from one set to another and the question therefore arises as to whether these differences are merely consequences of random error or whether the actual β's are different.

First, we shall change our subscripting so as to bring the notation of this section into line with that of Chapter 12, Section 9.4. Thus (w_{ij}, x_{ij}, Y_{ij}) represents the experimental data for the jth point of the ith set, $i = 1,2,\ldots r$; $j = 1,2,\ldots n_i$, so that there are r sets and n_i points in the ith set. We put $V(Y_{ij}) = \sigma_{ij}^2$, and $\sigma_i^2 = w_{ij}\sigma_{ij}^2$, independent of j. Similarly, B_{pi} stands for the estimator of β_{pi}, the coefficient of x^p in the expression for $E(Y_i)$, the expectation of Y in the ith set, viz.

$$E(Y_i) = \sum_p \beta_{pi} x^p .$$

We shall assume that $\sigma_i^2 = \sigma^2$ (unknown) for all i. This assumption may be tested, if necessary, by the methods of Chapter 9, Section 3, using the values $\{S_i^2\}$, $i = 1,2,\ldots r$, as mentioned in the first stage of the analogous tests given in Chapter 12.

Our two hypotheses are

$$H_0 : \beta_{p1} = \beta_{p2} = \ldots \beta_{pr} = \beta_p \text{ , say, all } p ,$$

$$H_1 : \text{not } H_0 ,$$

so that, under H_0, each B_{pi} is an estimator of the same quantity β_p; under the alternative hypothesis, each set of data is governed by a different polynomial with its own set of coefficients $\{\beta_{pi}\}$.

Now if H_0 is true, we may disregard the division of the data into sets and calculate a set of estimators $\{B_p\}$ from the combined data. Then, we may obtain a sum of squares

$$S_{DD,c} = \sum_i\sum_j w_{ij} D_{ij,c}^2 ,$$

where

$$D_{ij,c} = Y_{ij} - \sum_p B_p x_{ij}^p ,$$

and the c denotes that the data have been combined.

We also have sums of squares for the various sets; thus, for the ith set, we have

$$S_{DD,i} = \sum_j w_{ij} D_{ij}^2 ,$$

where

$$D_{ij} = Y_{ij} - \sum_p B_{pi} x_{ij}^p .$$

Putting

$$S_{DD} = \Sigma S_{DD,i} ,$$

it may be shown that, under H_O, $(S_{DD,c}-S_{DD})/\sigma^2$ and S_{DD}/σ^2, are independently distributed as χ^2 with $\{(n-m)-(n-rm)\} = m(r-1)$ and $(n-rm)$ degrees of freedom respectively, n here representing Σn_i. Thus $(S_{DD,c}-S_{DD})/\{m(r-1)\}$ and $S_{DD}/(n-rm)$ provide unbiased estimators of σ^2 under H_O. Hence, the test ratio

$$T = \frac{(S_{DD,c} - S_{DD})/\{m(r-1)\}}{S_{DD}/(n-rm)}$$

is distributed as $F_{m(r-1),n-rm}$ under the null hypothesis. Under the alternative hypothesis, the expectation of the numerator will be greater than that of the denominator, which provides an unbiased estimator of σ^2 in this case as well as under H_O. Consequently, we use a one-sided test, rejecting H_O at the α level of significance if $t \geq F_{m(r-1),n-rm}(1-\alpha)$.

If the data pass the test for homogeneity, that is if H_O is accepted, we use $\{B_p\}$ as our estimators of $\{\beta_p\}$ and $S_{DD,c}/(n-m)$ as our estimator of σ^2. An alternative estimator is $S_{DD}/(n-rm)$ which is unbiased under H_O or H_1. However, the estimator from the combined data is preferable when H_O is accepted.

If the data do not pass the test, the $\{B_{pi}\}$ afford estimators of the coefficients $\{\beta_{pi}\}$ and $S_{DD}/(n-rm)$ provides an estimator of σ^2.

If the β_{pi}'s are significantly different between sets, it may be premature to conclude that none of the sets have equal β's. Inspection of the coefficients $\{b_{pi}\}$ associated with the various sets may reveal that, in the case of one or more particular sets, the coefficients are greatly different from those associated with the remaining sets. In this situation, it is clearly worthwhile applying the test to the sets which look consistent in order to infer whether the failure of the original test is a consequence of 'rogue' data, which should then be confirmed using new data. Of course similar remarks could be made whenever a set of quantities are found to be significantly different.

10.4.1. Comparison of individual coefficients

It may happen that there is some reason to suppose that β_q, say, is the same for all sets. (The most likely situation is that all sets have a common tangent at the origin, so that $\beta_{1i} = \beta_1$, say, all i). If we write the null hypothesis as

$$H_0 : \beta_{q1} = \beta_{q2} = \ldots \beta_{qr} = \beta_q \text{ , say,}$$

and the alternative hypothesis as the complement, viz.

$$H_1 : \text{not } H_0 \text{ ,}$$

then a suitable test criterion is

$$T = \frac{\sum_i w(B_{qi})(B_{qi} - \bar{B}_q)^2/(r - 1)}{\sum_i\sum_j w_{ij}D_{ij}^2/(n - rm)} \text{ .}$$

Here $\{B_{qi}\}$ are the individual estimators of β_{qi} and $\bar{B}_q$ is their weighted mean, the weighting factors $w(B_{qi})$ being chosen in accordance with our usual rule, thus

$$w(B_{qi}) = \sigma^2/\sigma^2(B_{qi}) \text{ .}$$

Under H_0, the expectation of the numerator of T is σ^2 whereas under H_1, it will be greater because of the inequality of the β_{qi}. The denominator of T, which is $S_{DD}/(n-rm)$, is an unbiased estimator of σ^2 under both hypotheses and is distributed independently of the numerator. Hence we may test for the homogeneity of the set $\{\beta_{qi}\}$ by a simple one-sided F-test, rejecting H_0 at the α level of significance if $t \geq F_{r-1,n-rm}(1-\alpha)$.

Note that this test is the generalised test of the $\{\beta_{1i}\}$, described in more detail in Chapter 12. If the data pass the test, $\bar{B}_q$ is our estimator of β_q and its most convenient variance estimator is $S^2/\Sigma w(B_{qi})$, where S^2 is the denominator of T.

11. Consistency of future experiments

Suppose that we have examined a reasonably large number of sets of data for coincidence, that the data have passed the test, and that we have a set of coefficients $\{b_p\}$ calculated as in Section 10.4 from the combined data. If the number of data is sufficiently large, we may believe these estimates to be close enough to $\{\beta_p\}$ to be used in place of the true values for the purpose of assessing the consistency or otherwise of a set of coefficients obtained in a future experiment. By the same token, we may use the estimate s^2 as the value of σ^2. Using this approximation, we can devise a simple test of consistency.

The procedure utilizes the theory developed in Chapter 16, to which reference may be made. Briefly, the arguments are as follows. If $Y_1, Y_2, \ldots Y_n$ are a set of random variables with covariance matrix $\sigma^2\underline{V}$ such that $\underline{Y} \sim N(\underline{\mu}, \sigma^2\underline{V})$, then

$$\underline{Z} = \underline{U}^{-1}(\underline{Y} - \underline{\mu})/\sigma \sim N(\underline{O}, \underline{I}_n) ,$$

where $\underline{V} = \underline{U}\underline{U}'$. Here, $\underline{Z}$, $\underline{Y}$, and $\underline{\mu}$ are nx1 vectors and $\underline{V}$ is a nxn matrix. Clearly,

$$\underline{Z}'\underline{Z} = \sum_1^n z_i^2 ,$$

the sum of n squared standard normal variates, and hence

$$(\underline{Y} - \underline{\mu})'\underline{V}^{-1}(\underline{Y} - \underline{\mu})/\sigma^2 = \underline{Z}'\underline{Z} \sim \chi_n^2 .$$

In the present context, the random variables in question are the coefficients $\{B_i\}$, i = 1 to m, and $\sigma^2\underline{V}$ is the covariance matrix with elements given by equation (6.1.2). It is a fundamental assumption of our test that the σ^2 associated with the measurements under test is identical with the σ^2 associated with the original measurements. Then, under the hypothesis

$$H_0 : \underline{\beta}_f = \underline{\beta} ,$$

where $\underline{\beta}_f$ is the vector of parameters associated with our future experiment, it follows that

$$T = (\underline{B}_f - \underline{\beta})'\underline{V}^{-1}(\underline{B}_f - \underline{\beta})/\sigma^2 \sim \chi_m^2 .$$

Here $\underline{B}_f$ is the vector of estimators of $\underline{\beta}$ in our future experiment. Under the alternative hypothesis,

$$H_1 : \underline{\beta}_f \neq \underline{\beta} ,$$

the arguments of Chapter 16 show that T is distributed as non-central χ_m^2 with non-centrality parameter $(\underline{\beta}_f-\underline{\beta})'\underline{V}^{-1}(\underline{\beta}_f-\underline{\beta})/\sigma^2$.

Thus, to test some future experiment yielding a vector of estimates $\underline{b}_f$, we simply evaluate the test ratio

$$t = (\underline{b}_f - \underline{\beta})'\underline{V}^{-1}(\underline{b}_f - \underline{\beta})/\sigma^2 ,$$

replacing $\underline{\beta}$ by $\underline{b}_c$ and σ^2 by s_c^2, the variance estimate from our combined set of data. We reject H_0 if $t \geq \chi_m^2(1-\alpha)$ where α is some chosen significance level. In view of the approximations involved in using $\underline{b}_c$ in place of $\underline{\beta}$ and s_c^2 in place of σ^2 it will probably be wise to use a 1 percent significance level rather than the usual 5 percent, to avoid the possibility of labelling too high a proportion of future experiments as inconsistent with the basic consolidated set.

Two further points need to be made. First, the elements of $\underline{V}$ must be evaluated from the x values appropriate to the set under test and not from the x values used in the combined set; this is because the test criterion is concerned with the spread of the B's for the set under test. Second, if the new data pass the test, the estimates of $\underline{\beta}$ and σ^2 should be updated by combining the new data with the old.

We give an example of the use of this type of argument in Example 2 of Chapter 14.

Example 1. Here we illustrate the sampling distribution of the least squares estimators of the coefficients of a polynomial. For this purpose, we simulate an experimental situation using a computer to generate random numbers distributed as N(0,1), a procedure to which we have already referred in Chapters 6, 8 and 9.

We start with the expression

$$\mu = 3x + 2x^2 + x^3$$

and calculate the values of the set $\{\mu_i\}$, i = 1 to 10, for the x values shown in Table 13.2.

Table 13.2 Values of (x_i, μ_i) used in the simulation experiment

x_i	μ_i
-2.0	-6.000
-1.5	-3.375
-1.0	-2.000
-0.5	-1.125
-0.1	-0.281
0.1	0.321
0.5	2.125
1.0	6.000
1.5	12.375
2.0	22.000

Table 13.3 Least squares estimates of the coefficients of $\mu = 3x + 2x^2 + x^3$.

RUN	b_1	b_2	b_3	s^2
1	3.102	2.015	0.971	0.343
2	3.164	2.001	0.938	0.315
3	2.675	2.098	1.073	0.305
4	2.998	2.114	1.001	0.068
5	2.964	1.931	0.994	0.215
6	3.123	2.051	0.965	0.554
7	2.445	1.929	1.200	0.479
8	2.960	1.949	0.996	0.065
9	3.229	2.018	0.937	0.142
10	3.082	1.930	1.006	0.302
11	3.230	1.983	0.918	0.206
12	2.668	2.037	1.067	0.086
13	3.024	1.994	1.024	0.186
14	2.976	2.102	1.038	0.184
15	3.383	2.133	0.877	0.170
Means	3.002	2.019	1.000	0.241
Expectations	3.000	2.000	1.000	0.250

We then generate a set of realizations $\{y_i\}$ using the relation

$$(Y_i - \mu_i)/\sigma = \varepsilon ,$$

in which $\varepsilon \sim N(0,1)$. We have put $\sigma = 0.5$. From $\{y_i\}$, we compute the estimates b_1, b_2, b_3 and s^2 using the theory of this chapter, there being 10-3 = 7 degrees of freedom available. We have repeated this operation 15 times in all to obtain a sample of least squares estimates drawn from the population of the corresponding estimators, B_1, B_2, B_3 and S^2.

Our results are shown in Table 13.3. Also shown in this table are the means over the sample of the various estimates which, as can be seen, compare very well with the expectations. However, the table also reveals just how far particular estimates can be 'out' and the consequent danger of drawing too hard and fast conclusions from a single run.

Example 2. A problem which frequently arises is the construction and use of a calibration curve. Here, we exemplify the procedure by examining a set of data which were obtained in the course of calibrating a Pirani gauge for the measurement of pressure in a high vacuum system. The gauge reading was set to zero when the pressure in the system was zero, and thereafter a series of gauge readings were taken at various known pressures, P. The results are recorded in Table 13.4.

According to the investigator, the errors in reading the gauge were negligible in comparison to the errors of measurement of the pressure. Hence, we assume that the gauge readings are error-free. So, although we approximately 'fix' the value of the pressure and measure the corresponding gauge reading, it is appropriate to regard P, as a dependent variable and the gauge reading as the independent variable, x. We put

$$Y = \frac{10^3 P}{\text{torr}} \tag{1}$$

and assume

$$Y = \mu + \varepsilon , \tag{2}$$

with $\varepsilon \sim N(0,\sigma^2)$, for all x values, the ε's being independent.

Table 13.4 Calibration of a Pirani gauge

x = Reading of the Pirani gauge	$y = \frac{10^3 p}{\text{torr}}$
0	0
0.0238	0.0282
0.0858	0.0891
0.2563	0.2674
0.2736	0.2857
0.5330	0.5723
0.6371	0.6979
1.046	1.285
1.354	1.697
1.786	2.337
2.546	3.416
2.761	3.686
3.551	4.657
4.479	6.367
5.235	7.396
7.330	11.15
9.283	15.87

We now examine the data using the method of Section 10.3.2. to find the most appropriate order of polynomial to use. We fit successively polynomials of increasing order, testing, as we go on, the reduction in the sum of squares about the fitted curve caused by incrementing the order by one, against the estimate of σ^2 calculated from the residuals about the higher order polynomial. The procedure is shown in Table 13.5 together with the appropriate critical values of F for a one-sided test at the five percent level.

The results in the table clearly show significant reductions in the total sum of squares as the order of the polynomial is increased from 0 to 4. However, when the order is increased from 4 to 5, the reduction in the sum of squares is small compared to s^2, and our test criterion $T = R_5/(S_{DD,5}/11)$ is less than the critical value of $F_{5,11}(0.95) = 4.84$. Non-significant results are also obtained for higher orders, so we take 4 as the most appropriate order for the polynomial of our set of data. From the table, we see that the basic error of measurement of y is estimated

Table 13.5 Choosing the appropriate order of a polynomial with n = 16 data points

Order of polynomial m	Degrees of freedom $df = n-m$	Sum of squares (ss) $S_{DD,m}$	Estimate of σ^2 $s^2 = \frac{S_{DD,m}}{df}$	Reduction in ss $R_m = S_{DD,m-1} - S_{DD,m}$	Test ratio $T = \frac{R_m}{s^2}$	Critical value $F_{1,df}(0.95)$
0	16	529.3313	33.08	-	-	4.49
1	15	5.0622	0.3375	524.2691	1553.39	4.54
2	14	0.28040	0.02003	4.7818	238.73	4.60
3	13	0.12902	0.009925	0.15138	15.25	4.67
4	12	0.063185	0.005265	0.065835	12.50	4.75
5	11	0.062184	0.005653	0.001001	0.18	4.84
6	10	0.051348	0.005135	0.010836	2.11	4.96
7	9	0.045535	0.005059	0.005813	1.15	5.12

by $s = (5.265 \times 10^{-3})^{\frac{1}{2}} = 0.073$; hence the estimated standard deviation of our pressure observations is 7.3×10^{-5} torr.

The coefficients $\{a_p\}$ and $\{b_p\}$ of our polynomial together with their standard errors are shown in Table 13.6.

Table 13.6 The coefficients $\{a_p\}$ and $\{b_p\}$ for the 4th order polynomial together with their standard errors

p	a_p	$s(A_p)$	b_p	$s(B_p)$
1	1.5401	0.0049	1.108	0.060
2	0.0607	0.0020	0.124	0.037
3	0.00506	0.00094	-0.0200	0.0071
4	0.00147	0.00041	0.00147	0.00041

We may thus write the equation of the fitted curve either as a polynomial in x, thus

$$\hat{\mu} = 1.108x + 0.124x^2 - 0.0200x^3 + 0.00147x^4 , \quad (3)$$

or as a linear combination of the $T_i(x)$, $i = 1$ to 4, thus

$$\hat{\mu} = 1.5401T_1(x) + 0.0607T_2(x) + 0.00506T_3(x) + 0.00147T_4(x) . \quad (4)$$

A plot of the fitted curve and the experimental points is shown in Figure 13.1. It is not possible on a figure of this scale to gain much idea of the accuracy of the polynomial approximation to the data. For this reason, we present in Table 13.7 a comparison of the values of y_i and $\hat{\mu}_i$, $i = 1$ to 16.

In considering the fit of this particular polynomial to our data, we also examine the first and second derivatives of equation (3). We find that the polynomial has a minimum value at $x = -2.53$ and crosses the x axis again at $x = -4.53$. There are no maxima or minima and no points of inflexion in the region of interest $(0 \leq x \leq 15.87)$. Consequently we have a polynomial which behaves as we should expect it to do on general physical grounds over the range of x values studied in the experiment.

Table 13.7 Comparison of $\{y_i\}$ and $\{\hat{\mu}_i\}$

y_i	$\hat{\mu}_i$	$y_i - \hat{\mu}_i$
0.0282	0.02644	+0.00176
0.0891	0.09595	-0.00685
0.2674	0.29174	-0.02434
0.2857	0.31198	-0.02628
0.5723	0.62279	-0.05049
0.6979	0.75120	-0.05330
1.285	1.2734	+0.0116
1.697	1.6827	+0.0143
2.337	2.2753	+0.0617
3.416	3.3565	+0.0595
3.686	3.6690	+0.0170
4.657	4.8364	-0.1794
6.367	6.2449	+0.1221
7.396	7.4329	-0.0369
11.15	11.146	+0.004
15.87	15.871	-0.001

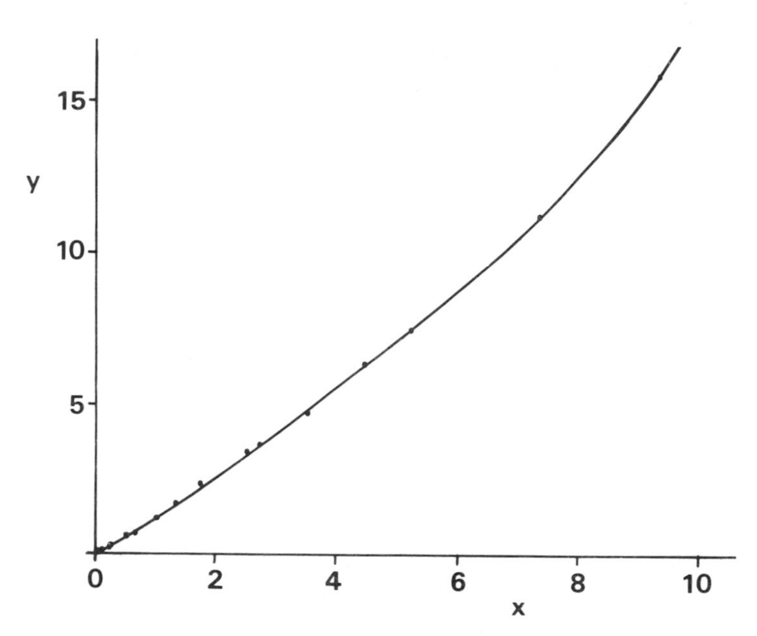

Figure 13.1. The fitted quartic (solid line) through the experimental points.

Thus, we may use equations (3) or (4) for interpolation purposes with some confidence. For example, let us calculate the estimate of μ at $x = 2.000$. We shall use both equations (3) and (4), listing the consecutive contributions of the various terms for the sake of interest:

$$\text{Equation (3): } \hat{\mu}_{2.000} = 2.216 + 0.496 - 0.160 + 0.024 = 2.576 \quad ,$$

$$\text{Equation (4): } \hat{\mu}_{2.000} = 3.080 - 0.592 + 0.110 - 0.024 = 2.574 \quad .$$

The small discrepancy has arisen entirely as a result of rounding-off error; in fact working to five decimal places, we get $\hat{\mu} = 2.57548$ [equation (3)] and $\hat{\mu} = 2.57550$ [equation (4)].

The values of $\{T_p(2.000)\}$ substituted into equation (4) were calculated from the expressions:

$$\begin{aligned} T_1(x) &= x \ , \\ T_2(x) &= -6.8730x + x^2 \ , \\ T_3(x) &= 31.519x - 12.314x^2 + x^3 \ , \\ T_4(x) &= -119.22x + 85.766x^2 - 17.089x^3 + x^4 \ , \end{aligned}$$

the various coefficients being taken from Table 13.8 which, in addition, gives the values of g_{pq} (see Appendix 2) and some elements of the cross-check matrix $\sum_i T_p(x_i)T_q(x_i)$, for p,q = 1,2,...7. We obtained

$$T_1(2.000) = 2.000$$

$$T_2(2.000) = -9.746\ ,$$

$$T_3(2.000) = 21.78\ ,$$

$$T_4(2.000) = -16.09\ .$$

Finally, we calculate $s^2(\hat{\mu})$ using equation (5.4.9), viz.

$$s^2(\hat{\mu}) = s^2 \sum_p \frac{T_p^2(x)}{\sum_i T_p^2(x_i)}\ ,$$

since $w_i = 1$, all i. We read off the values of $\sum_i T_p^2(x_i)$ which we require from the diagonal of the cross-check matrix (see Table 13.8). This matrix thus provides not only a check on the orthogonality of the T_p's at the data points (the off-diagonal elements only differ from zero as a result of rounding-off error in the computer employed to make the calculations), but also data for the computation of the error of interpolation. Our calculations proceed as follows:

p	$\sum_i T_p^2(x_i)$	$T_p^2(2.000)$	Contribution to $\sum_p$
1	221.04	4.000	0.01810
2	1298.9	94.98	0.07312
3	5912.2	474.4	0.08024
4	30637	258.9	0.00845
			0.17991

The value of s^2 is 5.265×10^{-3} (from Table 13.5). Hence

$$s^2(\hat{\mu}) = 5.265 \times 10^{-3} \times 0.17991$$

$$= 9.47 \times 10^{-4}\ ,$$

Table 13.8 (i) The coefficients g_{pq}

6.8730	1					
5.8763	5.4409	1				
	4.5517	4.7750	1			
		5.1820	4.5250	1		
			3.7345	3.3614	1	
				1.8346	2.7479	1

(ii) The coefficients c_{pq}

1						
-6.8730	1					
31.519	-12.314	1				
-119.22	85.766	-17.089	1			
376.13	-443.50	157.91	-21.614	1		
-819.10	1546.6	-910.48	226.83	-24.975	1	
1560.7	-4255.3	3758.8	-1494.1	293.62	-27.723	1

(iii) The diagonal elements of the cross-check matrix, viz. $\sum_i w_i T_p^2(x_i)$, p = 1 to 7

p	1	2	3	4	5	6	7
$\sum_i T_p^2(x_i)$	2.2104×10^2	1.2989×10^3	5.9122×10^3	3.0637×10^4	1.1441×10^5	2.0991×10^5	4.2042×10^5

Notes (a) Where no entry in the grid is shown, the value of the corresponding element is zero.

(b) The order of magnitude of the off-diagonal elements of the cross-check matrix was 10^{-5} or less.

giving

$$s(\hat{\mu}) = 0.031 .$$

Reverting back to pressure, we conclude that the pressure corresponding to a gauge reading of 2.000 is 2.575×10^{-3} torr with a standard error of 3.1×10^{-5} torr based on 12 degrees of freedom.

CHAPTER 14

THE GENERAL STRAIGHT LINE

The problem which we discuss here is the question of fitting a straight line to a set of data $\{w_i, x_i, y_i\}$. Unlike the case treated in Chapter 12, the fitted line is not constrained to pass through the origin or through some other fixed point. This problem occurs frequently in many experimental situations and for this reason we devote a special chapter to its solution.

We follow a similar pattern to that of the previous two chapters. Firstly, we require a clear statement of the model for the experimental data, a model which should preferably be backed by some theoretical reasoning about the relationship between the variables x and $E(Y) = \mu$. Secondly, we shall show how to set up estimators of the parameters associated with the model. Finally, we shall show how to compute confidence intervals and test a number of hypotheses on the basis of an assumed distribution for the error term, including testing whether two or more fitted lines are 'really' coincident or parallel or have a common intercept.

As in the previous two chapters, we assume that the Y's are independent.

1. Formal statement of the model

We shall assume that the random variable Y is related to x by the equation

$$Y = \beta_0 + \beta_1 x + \varepsilon , \qquad (1)$$

where β_0 and β_1 are constants and ε is the error term with the properties:

$$E(\varepsilon) = 0 \text{ for all } x, \qquad (2)$$

$$V(\varepsilon) = \sigma_i^2 \text{ for } x = x_i \; . \qquad (3)$$

Hence

$$E(Y) = \mu = \beta_0 + \beta_1 x , \tag{4}$$

$$V(Y_i) = \sigma_i^2 . \tag{5}$$

We shall follow our previous practice of formulating our expressions in terms of a set of known weights, $\{w_i\}$, and an unknown constant, σ^2, where

$$w_i = \sigma^2/\sigma_i^2 . \tag{6}$$

The special case where all the experimental points are associated with the same variance is simply dealt with by putting $w_i = 1$, all i.

2. Estimation using the method of least squares

As before, we seek the estimators B_0 and B_1 of β_0 and β_1 which minimise the sum of squares

$$\$ = \Sigma w_i (Y_i - \mu_i)^2 , \tag{1}$$

where

$$\mu_i = E(Y_i) = \beta_0 + \beta_1 x_i . \tag{2}$$

The usual procedure gives two simultaneous equations for B_0 and B_1, the solutions of which are

$$B_0 = \frac{(\Sigma w_i x_i^2)(\Sigma w_i Y_i) - (\Sigma w_i x_i)(\Sigma w_i x_i Y_i)}{(\Sigma w_i)(\Sigma w_i x_i^2) - (\Sigma w_i x_i)^2} , \tag{3}$$

$$B_1 = \frac{(\Sigma w_i)(\Sigma w_i x_i Y_i) - (\Sigma w_i x_i)(\Sigma w_i Y_i)}{(\Sigma w_i)(\Sigma w_i x_i^2) - (\Sigma w_i x_i)^2} . \tag{4}$$

Note that the denominator is the same in these last two equations. Alternative forms of equations (3) and (4) are

$$B_0 = \bar{Y} - B_1 \bar{x} , \tag{5}$$

$$B_1 = \frac{\Sigma w_i Y_i (x_i - \bar{x})}{\Sigma w_i (x_i - \bar{x})^2} \tag{6}$$

$$= \frac{\Sigma w_i x_i Y_i - (\Sigma w_i Y_i)(\Sigma w_i x_i)/\Sigma w_i}{\Sigma w_i x_i^2 - (\Sigma w_i x_i)^2/\Sigma w_i} , \tag{7}$$

where $\bar{x}$, $\bar{Y}$ are the weighted means of the x_i, Y_i viz.

$$\bar{x} = \Sigma w_i x_i / \Sigma w_i , \tag{8}$$

$$\bar{Y} = \Sigma w_i Y_i / \Sigma w_i . \tag{9}$$

Equation (6) is similar to equation (1) of Chapter 12, Section 6.

The equation of the fitted line is thus

$$\hat{\mu} = B_0 + B_1 x . \tag{10}$$

If equations (3) and (4) are used, it is advisable to check the dimensions to make sure that the expressions for the slope and intercept have not been interchanged. However, equations (7) and (5) are preferable for computation.

When each of the weighting factors is unity, that is, when $V(Y_i) = \sigma^2$, all i, equations (3) to (6) reduce to the simple expressions

$$B_0 = \frac{(\Sigma x_i^2)(\Sigma Y_i) - (\Sigma x_i)(\Sigma x_i Y_i)}{n\Sigma x_i^2 - (\Sigma x_i)^2}$$

$$= \bar{Y} - B_1 \bar{x} ,$$

$$B_1 = \frac{n\Sigma x_i Y_i - (\Sigma x_i)(\Sigma Y_i)}{n\Sigma x_i^2 - (\Sigma x_i)^2}$$

$$= \frac{\Sigma Y_i (x_i - \bar{x})}{\Sigma (x_i - \bar{x})^2}$$

$$= \frac{\Sigma x_i Y_i - (\Sigma x_i)(\Sigma Y_i)/n}{\Sigma x_i^2 - (\Sigma x_i)^2/n} ,$$

where

$$\bar{x} = \Sigma x_i/n \quad \text{and} \quad \bar{Y} = \Sigma Y_i/n .$$

3. Estimation using the method of least squares and polynomials $T_0(x)$ and $T_1(x)$

In this section we introduce two orthogonal polynomials in x (see Appendix 2) which provide an alternative algebraic means of solving the estimation problem under discussion. These polynomials are defined by the relations

$$T_0(x) = 1 , \qquad (1)$$

$$T_1(x) = f_{10} + x . \qquad (2)$$

The quantity f_{10} is fixed by making these two polynomials orthogonal at the data points; that is, f_{10} is chosen so as to give

$$\Sigma w_i T_0(x_i) T_1(x_i) = 0 . \qquad (3)$$

Obviously,

$$f_{10} = -\Sigma w_i x_i / \Sigma w_i = -\bar{x} .$$

Hence, explicit expressions for our two polynomials are

$$T_0(x) = 1 ,$$

$$T_1(x) = x - \bar{x} .$$

We recast our model for the random variable Y in terms of these polynomials, thus

$$Y = \alpha_0 T_0(x) + \alpha_1 T_1(x) + \varepsilon , \qquad (4)$$

where the error term has the properties defined earlier in

equations (1.2) and (1.3). Hence

$$E(Y) = \mu = \alpha_0 T_0(x) + \alpha_1 T_1(x) . \tag{5}$$

We use the method of least squares to find estimators A_0 and A_1 of α_0 and α_1. Using the orthogonality property at the data points, it is a relatively easy matter to show that

$$A_0 = \frac{\Sigma w_i Y_i T_0(x_i)}{\Sigma w_i T_0^2(x_i)} , \tag{6}$$

$$A_1 = \frac{\Sigma w_i Y_i T_1(x_i)}{\Sigma w_i T_1^2(x_i)} . \tag{7}$$

We draw attention to the formal similarity between the two results (c.f. B_0 and B_1). Analogous expressions can be obtained relating α_0 and α_1 to $\{\mu_i\}$ by the use of the orthogonality condition:

$$\alpha_0 = \frac{\Sigma w_i \mu_i T_0(x_i)}{\Sigma w_i T_0^2(x_i)} , \tag{8}$$

$$\alpha_1 = \frac{\Sigma w_i \mu_i T_1(x_i)}{\Sigma w_i T_1^2(x_i)} . \tag{9}$$

Since $T_0(x_i) = 1$, all i, and $T_1(x_i) = x_i - \bar{x}$, alternative expressions for A_0 and A_1 are

$$A_0 = \Sigma w_i Y_i / \Sigma w_i = \bar{Y} , \tag{10}$$

$$A_1 = \frac{\Sigma w_i Y_i (x_i - \bar{x})}{\Sigma w_i (x_i - \bar{x})^2} = B_1 . \tag{11}$$

It is sometimes useful to include $\bar{Y}$ in the expression for A_1. Since $\Sigma w_i \bar{Y}(x_i - \bar{x}) = 0$, equation (11) may be written

$$A_1 = \frac{\Sigma w_i (Y_i - \bar{Y})(x_i - \bar{x})}{\Sigma w_i (x_i - \bar{x})^2} . \quad (12)$$

The equation of the fitted line is

$$\hat{\mu} = A_0 T_0(x) + A_1 T_1(x) . \quad (13)$$

An alternative form of equation (13) is

$$\hat{\mu} = \bar{Y} + A_1(x - \bar{x}) . \quad (14)$$

This representation is not only simpler mathematically than that given by equation (2.10), but also is often more appropriate in physical terms (for instance, see Example 1). Equation (14) shows that the fitted line passes through $(\bar{x},\bar{y})$, the centroid of the data points. Since $\bar{x}$ and $\bar{y}$ are very easily calculated, a quick method of approximately locating the least squares line through a particular set of data points is to mark in the centroid on the graph and then draw the best 'eye-line' through this which is consistent with the experimental values.

3.1. Expectations of the estimators

Since A_0, A_1, B_0, and B_1 are linear combinations of the $\{Y_i\}$, it is easy to show that

$$E(A_0) = \alpha_0 , \qquad E(A_1) = \alpha_1 ,$$

$$E(B_0) = \beta_0 , \qquad E(B_1) = \beta_1 ,$$

so that A_0, A_1, B_0, and B_1 are unbiased.

From either of these two sets of results it follows that

$$E(\hat{\mu}) = \mu ,$$

showing that the estimators of μ are unbiased as well.

3.2. The covariance matrix of (A_0,A_1) and the variance of $\hat{\mu}$

The arguments of Chapter 13, Section 5.2 show that

$$C(A_p,A_q) = \frac{\sigma^2 \sum_i w_i T_p(x_i) T_q(x_i)}{\left\{\sum_i w_i T_p^2(x_i)\right\}\left\{\sum_i w_i T_q^2(x_i)\right\}} . \tag{1}$$

Hence, using the orthogonality property,

$$C(A_0,A_1) = 0 .$$

Furthermore, since $T_0(x_i) = 1$ and $T_1(x_i) = x_i - \bar{x}$,

$$V(A_0) = \sigma^2/\Sigma w_i , \tag{2}$$

$$V(A_1) = \frac{\sigma^2}{\Sigma w_i (x_i - \bar{x})^2} \tag{3}$$

$$= \frac{\sigma^2}{\Sigma w_i x_i^2 - (\Sigma w_i x_i)^2/\Sigma w_i} . \tag{4}$$

To find $V(\hat{\mu})$, we write

$$\hat{\mu} = A_0 T_0(x) + A_1 T_1(x) .$$

Since $C(A_0,A_1) = 0$, it follows at once that

$$V(\hat{\mu}) = T_0^2(x) V(A_0) + T_1^2(x) V(A_1)$$

$$= \sigma^2 \left\{\frac{1}{\Sigma w_i} + \frac{(x - \bar{x})^2}{\Sigma w_i (x_i - \bar{x})^2}\right\} . \tag{5}$$

Equation (5) shows that the minimum variance of $\hat{\mu}$ occurs at $x = \bar{x}$. The graph of the standard deviation of $\hat{\mu}$ against x is illustrated in Figure 14.1. We shall return to this point when we discuss interpolation and extrapolation using the fitted line.

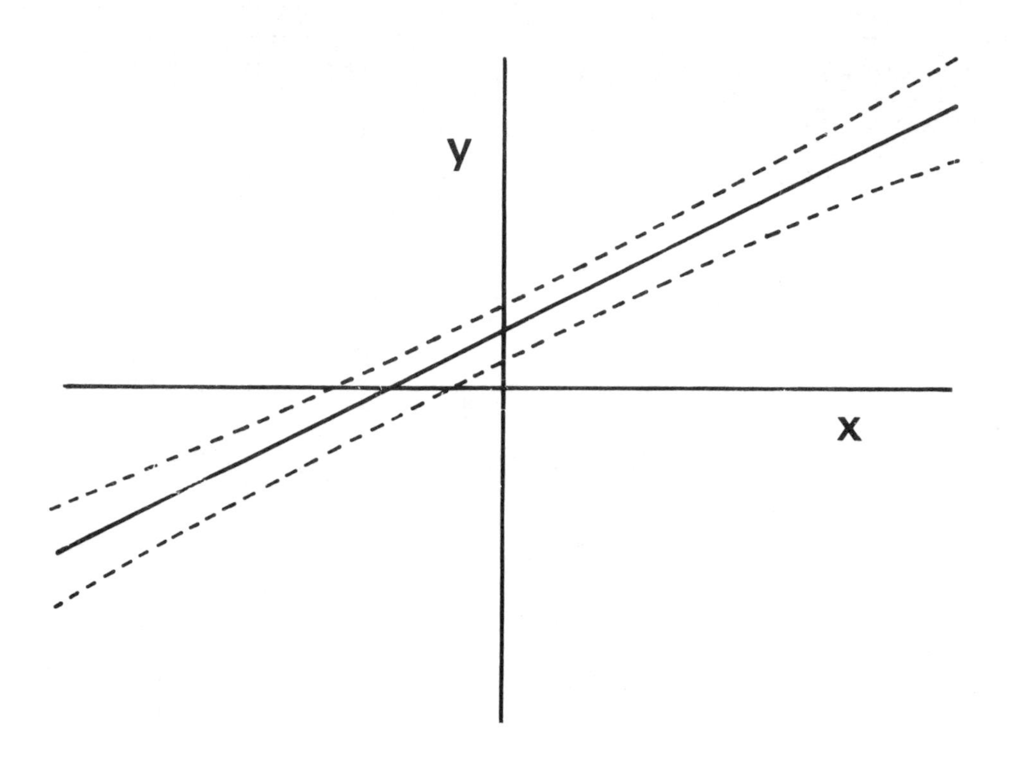

Figure 14.1. The variation of $\sigma(\hat{\mu})$ with x about the line $\hat{\mu} = \bar{Y} + A_1(x - \bar{x})$

We may also use equation (5) to find the expression for $V(B_0)$. Since $B_0 = \hat{\mu}$ when $x = 0$, we obtain

$$V(B_0) = \sigma^2\left\{\frac{1}{\Sigma w_i} + \frac{\bar{x}^2}{\Sigma w_i(x_i - \bar{x})^2}\right\}$$

$$= \frac{\sigma^2 \Sigma w_i x_i^2}{\Sigma w_i \; \Sigma w_i(x_i - \bar{x})^2} \quad . \tag{6}$$

Since $B_1 = A_1$, $V(B_1)$ is given by equation (3) or (4).

3.3. Formulae for covariances

We require $C(\hat{\mu},A_1)$. Now

$$\begin{aligned} C(\hat{\mu},A_1) &= E\{(\hat{\mu} - \mu)(A_1 - \alpha_1)\} \\ &= E[\{(A_0-\alpha_0)T_0(x) + (A_1-\alpha_1)T_1(x)\}(A_1-\alpha_1)] \\ &= T_0(x)C(A_0,A_1) + T_1(x)V(A_1) \\ &= T_1(x)V(A_1) \\ &= \frac{\sigma^2(x - \bar{x})}{\Sigma w_i(x_i - \bar{x})^2} \,. \end{aligned} \tag{1}$$

A particular case of importance corresponds to $x = 0$ where $\hat{\mu} = B_0$. Then, since $B_1 = A_1$, we obtain

$$C(B_0,B_1) = \frac{-\bar{x}\sigma^2}{\Sigma w_i(x_i - \bar{x})^2} \,. \tag{2}$$

Combining the expressions for $V(B_0)$ and $V(B_1)$ with equation (2), we may easily show that

$$\rho(B_0,B_1) = -\bar{x}/(\overline{x^2})^{\frac{1}{2}} \,, \tag{3}$$

where $\overline{x^2} = \Sigma w_i x_i^2/\Sigma w_i$.

3.4. The weighted sum of squared residuals and derived variance estimators

We have seen in Chapter 13, Section 6.3. that the sum of squared residuals is

$$\Sigma w_i D_i^2 = \Sigma w_i Y_i^2 - \Sigma w_i \hat{\mu}_i^2 \,, \tag{1}$$

which becomes in this case

$$\left.\begin{aligned} \Sigma w_i D_i^2 &= \Sigma w_i Y_i^2 - A_0^2 \Sigma w_i T_0^2(x_i) - A_1^2 \Sigma w_i T_1^2(x_i) \\ &= \Sigma w_i (Y_i - \bar{Y})^2 - A_1^2 \Sigma w_i T_1^2(x_i) \,, \end{aligned}\right\} \tag{2}$$

because $A_0 = \bar{Y}$ and $T_0(x_i) = 1$, all i. If we replace A_1 by equation (3.12) and $T_1(x_i)$ by $(x_i - \bar{x})$ we obtain

$$\Sigma w_i D_i^2 = \Sigma w_i (Y_i - \bar{Y})^2 - \frac{\{\Sigma w_i (Y_i - \bar{Y})(x_i - \bar{x})\}^2}{\Sigma w_i (x_i - \bar{x})^2} . \quad (3)$$

An alternative form of equation (3), which is more convenient for computation, is

$$\Sigma w_i D_i^2 = S'_{YY} - S'^2_{xY}/S'_{xx} ,$$

where

$$S'_{YY} = \Sigma w_i (Y_i - \bar{Y})^2$$

$$= S_{YY} - S_Y^2/S_w ,$$

$$S'_{xY} = \Sigma w_i (x_i - \bar{x})(Y_i - \bar{Y})$$

$$= S_{xY} - S_x S_Y/S_w ,$$

$$S'_{xx} = \Sigma w_i (x_i - \bar{x})^2$$

$$= S_{xx} - S_x^2/S_w ,$$

with $S_{YY} = \Sigma w_i Y_i^2$, $S_{xY} = \Sigma w_i x_i Y_i$, and $S_{xx} = \Sigma w_i x_i^2$.

We recognize the first term on the right hand side of equation (3) as the weighted sum of squared displacements of the Y's from their mean, the total sum of squares as we shall call it. We also understand that $\Sigma w_i D_i^2$ is the residual sum of squares which remains once account has been taken of the variation of Y with x. Hence the difference

$$\Sigma w_i (Y_i - \bar{Y})^2 - \Sigma w_i D_i^2 = \frac{\{\Sigma w_i (Y_i - \bar{Y})(x_i - \bar{x})\}^2}{\Sigma w_i (x_i - \bar{x})^2}$$

is that part of the total sum of squares which is 'accounted for' or 'explained by' or 'due to' the regression. Thus the proportion of the total sum of squares accounted for by the regression is

$$\frac{\{\Sigma w_i(Y_i - \bar{Y})(x_i - \bar{x})\}^2}{\Sigma w_i(Y_i - \bar{Y})^2 \Sigma w_i(x_i - \bar{x})^2} \, . \tag{4}$$

Now this expression is the square of the sample correlation coefficient for the general case where the data are weighted. (If $w_i = 1$, all i, we recover the form given in Chapter 11). Thus r^2 is the proportion of the variation of Y accounted for by its regression on x. Since r is symmetrical in x and Y, a similar statement would be true if the roles of x and Y were interchanged.

We now turn to the expectation of $\Sigma w_i D_i^2$. Using equation (1), we have

$$\begin{aligned} E(\Sigma w_i D_i^2) &= \Sigma w_i E(Y_i^2) - \Sigma w_i E(\hat{\mu}_i^2) \\ &= \Sigma w_i V(Y_i) - \Sigma w_i V(\hat{\mu}_i) \\ &= n\sigma^2 - V(A_0)\Sigma w_i T_0^2(x_i) - V(A_1)\Sigma w_i T_1^2(x_i) \\ &= (n - 2)\sigma^2 \, . \end{aligned}$$

Hence

$$s^2 = \Sigma w_i D_i^2/(n - 2) \tag{5}$$

is an unbiased estimator of σ^2. The number of degrees of freedom, n-2, is again the number of points, n, minus the number of estimated parameters. Using s^2 we may write down equations for the estimators of variance and covariance. These are

$$s^2(A_0) = s^2/\Sigma w_i \, , \tag{6}$$

$$s^2(B_0) = s^2\left\{\frac{1}{\Sigma w_i} + \frac{\bar{x}^2}{\Sigma w_i(x_i - \bar{x})^2}\right\} , \tag{7}$$

$$s^2(A_1) = s^2(B_1) = \frac{s^2}{\Sigma w_i(x_i - \bar{x})^2} \, , \tag{8}$$

$$s^2(\hat{\mu}) = s^2\left\{\frac{1}{\Sigma w_i} + \frac{(x - \bar{x})^2}{\Sigma w_i(x_i - \bar{x})^2}\right\}, \tag{9}$$

$$\hat{C}(B_0,B_1) = \frac{-s^2\bar{x}}{\Sigma w_i(x_i - \bar{x})^2}. \tag{10}$$

We shall find this last result useful when we come to estimate the standard deviation of a quantity calculated from both the slope and intercept of a fitted line.

3.5. A note concerning rounding-off errors

When calculating weighted sums of squares or cross products about the means like $\Sigma w_i(x_i - \bar{x})^2$, $\Sigma w_i(x_i - \bar{x})(y_i - \bar{y})$, which may be written as $S'_{xx} = \Sigma w_i x_i^2 - (\Sigma w_i x_i)^2/\Sigma w_i$, $S'_{xy} = \Sigma w_i x_i y_i - (\Sigma w_i x_i)(\Sigma w_i y_i)/\Sigma w_i$, we must beware of a drastic loss in accuracy which can occur when the coefficient of variation of the x's or of the y's is small. In particular when the x's and/or the y's have a few leading significant figures which are virtually constant throughout the data, then more leading significant figures in quantities like $\Sigma w_i x_i^2$, $\Sigma w_i x_i y_i$, $(\Sigma w_i x_i)^2/\Sigma w_i$ etc. will be virtually constant, and these will largely cancel out in calculating S'_{xx}, S'_{xy}, etc., leaving far less accuracy in these quantities than might be supposed. Working with a desk machine, one can usually see how much accuracy is being retained, while this is often not obvious when using a computer. A widely used technique to allow for this is given by Welford (1962). Alternatively, one may simply subtract a value roughly equal to the mean $\bar{x}$ from each x value (it could even by the first x value to save assessing the mean), and work with these coded data - similarly for the y's. Obviously this transformation does not change S'_{xx}, S'_{xy}, etc. If $\bar{x}$ and/or $\bar{y}$ are to be obtained in the same set of calculations, and x', y' represent the coded data, so that

$$x' = x - a,$$

$$y' = y - b,$$

then clearly

$$\bar{x} = \bar{x}' + a ,$$

$$\bar{y} = \bar{y}' + b .$$

There is a related discussion in Chapter 5, Section 4.1.

3.6. Interpolation and extrapolation using the fitted line

The relevant equations are

$$\hat{\mu} = B_0 + B_1 x = \bar{Y} + A_1(x - \bar{x}) \quad (1)$$

and

$$S(\hat{\mu}) = S\left\{\frac{1}{\Sigma w_i} + \frac{(x - \bar{x})^2}{\Sigma w_i (x_i - \bar{x})^2}\right\}^{\frac{1}{2}} . \quad (2)$$

Clearly the graph of $\hat{\mu} \pm S(\hat{\mu})$ against x will be similar to Figure 14.1. Equation (2) shows that the location of the estimator $\hat{\mu}$ becomes more uncertain the further away x is from $\bar{x}$. For inverse interpolation where we wish to estimate x corresponding to an observation Y, of weight w, or to a given value of μ, we use

$$\hat{x}_Y = \bar{x} + (Y - A_0)/A_1, \quad \text{or} \quad (3)$$

$$\hat{x}_\mu = \bar{x} + (\mu - A_0)/A_1 . \quad (4)$$

Using the method of Chapter 5, Section 5, we have $E(\hat{x}_Y) \simeq E(\hat{x}_\mu) \simeq x$, so that the $\hat{x}$'s are approximately unbiased. Also

$$S^2(\hat{x}_Y) = S^2[A_1^2/w + A_1^2/\Sigma w_i + (Y-A_0)^2/\{A_1^4 \Sigma w_i (x_i-\bar{x})^2\}]$$

$$= S^2\{1/w + 1/S_w + (\hat{x}_Y-\bar{x})^2/S'_{xx}\}/A_1^2 . \quad (5)$$

Similarly,

$$S^2(\hat{x}_\mu) = S^2\{1/S_w + (\hat{x}_\mu-\bar{x})^2/S'_{xx}\}/A_1^2 . \quad (6)$$

4. Sampling distributions assuming the normality of the data

When the Y's are independently normally distributed, a number of important consequences follow.

Firstly, the estimators A_0, A_1, B_0 and B_1 are maximum likelihood estimators possessing all the useful properties of such. Secondly, because each of these estimators and also $\hat{\mu}$ is a linear combination of the $\{Y_i\}$, each is itself normally distributed. Hence, we may write

$$A_0 \sim N(\alpha_0, \sigma^2/\Sigma w_i) \ , \tag{1}$$

$$A_1 \sim N\{\alpha_1, \sigma^2/\Sigma w_i(x_i - \bar{x})^2\}, \tag{2}$$

$$B_0 \sim N[\beta_0, \sigma^2\{1/\Sigma w_i + \bar{x}^2/\Sigma w_i(x_i - \bar{x})^2\}] \ , \tag{3}$$

$$B_1 \sim N\{\beta_1, \sigma^2/\Sigma w_i(x_i - \bar{x})^2 \} \ , \tag{4}$$

$$\hat{\mu} \sim N[\mu, \sigma^2\{1/\Sigma w_i + (x - \bar{x})^2/\Sigma w_i(x_i - \bar{x})^2\}] \ . \tag{5}$$

Thirdly, it may be shown that

$$\Sigma w_i D_i^2/\sigma^2 = (n-2)S^2/\sigma^2 \sim \chi^2_{n-2} \ , \tag{6}$$

independently of A_0, A_1, B_0, B_1 and $\hat{\mu}$. Fourthly, because $C(A_0, A_1) = 0$, the two estimators, A_0 and A_1, are independent.

5. Calculation of confidence intervals for $\{\alpha_p\}$, $\{\beta_p\}$, and μ

Under the assumption of normality of the $\{Y_i\}$, we can easily calculate confidence intervals for the parameters α_0, α_1, β_0, β_1 and μ. We illustrate the procedure for the case of β_0. Since

$$(B_0 - \beta_0)/\sigma(B_0) \sim N(0,1) \ , \tag{1}$$

and

$$(n-2)S^2(B_0)/\sigma^2(B_0) = (n-2)S^2/\sigma^2 \sim \chi^2_{n-2} \ , \tag{2}$$

independently of B_0, it follows that

$$(B_0 - \beta_0)/S(B_0) \sim t_{n-2} \,. \tag{3}$$

Thus the confidence interval for β_0 with coefficient $1-\alpha$ is

$$b_0 \pm t_{n-2}(1-\alpha/2)\,s(B_0) \;, \tag{4}$$

where b_0 is the particular realization of B_0.

6. Tests of a number of hypotheses

In this section, we consider a number of tests relating to β_0, β_1, and μ. Any tests on β_1 are, of course, applicable to α_1. However, we shall not consider any tests which might be made on α_0 since such tests are unlikely to have any real value.

6.1. Tests on β_0

The usual situation is that we wish to test the hypotheses

$$H_0 : \beta_0 = \beta_{0,0} \quad \begin{cases} H_1 : \beta_0 > \beta_{0,0} \,, \\ H_1 : \beta_0 < \beta_{0,0} \,, \quad \text{or} \\ H_1 : \beta_0 \neq \beta_{0,0} \,. \end{cases}$$

Here $\beta_{0,0}$ is some preconceived value of β_0, from theory, or perhaps from the literature assuming the value quoted is error-free; often $\beta_{0,0}$ will be zero.

Under H_0, the statistic $(B_0 - \beta_{0,0})/S(B_0)$ will be distributed as t_{n-2} and so we may decide whether or not to accept H_0 in precisely the same way as in Chapter 8, Section 1.

If H_0 states that $\beta_{0,0} = 0$ and the result of the test is that H_0 is accepted, we should use $\Sigma w_i x_i Y_i / \Sigma w_i x_i^2 = S_{xY}/S_{xx}$ as our estimator of β_1 (as in Chapter 12). Under H_0, this is preferable to S'_{xY}/S'_{xx} although the latter is an unbiased estimator of β_1 whether or not H_0 is true.

6.2. Tests on β_1

We formulate our two hypotheses as in Section 6.1.

$$H_0 : \beta_1 = \beta_{1,0} \quad \left| \begin{array}{l} H_1 : \beta_1 > \beta_{1,0} , \\ H_1 : \beta_1 < \beta_{1,0} , \text{ or} \\ H_1 : \beta_1 \neq \beta_{1,0} . \end{array} \right.$$

Under H_0, $(B_1 - \beta_{1,0})/S(B_1) \sim t_{n-2}$ and so we may decide upon the acceptance/rejection of H_0 in the usual way.

When our null hypothetical value, $\beta_{1,0}$, is zero, this test amounts to testing whether or not there is any significant dependence of Y on x. Thus, the test is closely related to the test of significance of r, the sample correlation coefficient, although in the latter case both X and Y are random variables. Indeed the two tests are exactly equivalent. With

$$r = \frac{\Sigma w_i (X_i - \bar{X})(Y_i - \bar{Y})}{\{\Sigma w_i (X_i - \bar{X})^2 \Sigma w_i (Y_i - \bar{Y})^2\}^{\frac{1}{2}}} = \frac{S'_{XY}}{(S'_{XX} S'_{YY})^{\frac{1}{2}}} ,$$

where

$$\bar{X} = S_X/S_w , \quad \bar{Y} = S_Y/S_w ,$$

and X_i, Y_i jointly normally distributed with variances σ_1^2/w_i and σ_2^2/w_i and zero distribution correlation coefficient, ρ, then

$$T = r\left(\frac{n-2}{1-r^2}\right)^{\frac{1}{2}} \sim t_{n-2} .$$

It can easily be shown that this test criterion is equal to $B_1/S(B_1)$, with X replaced by x.

6.3. Test on μ

This is carried out similarly to the test on β_0.

6.4. Test of goodness of fit of the model

The procedure here is precisely analogous to that explained in Chapters 12 and 13. Consequently we need do little more than outline the test.

We assume that we have available r sets of replicate observations $\{Y_{ij}\}$ $i = 1,2,\ldots r$; $j = 1,2,\ldots n_i$. Under the model $Y_{ij} \sim N(\sum_{p=0}^{1}\beta_p x_i^p, \sigma^2/w_i)$, where the $\{\beta_p\}$ and σ^2 are unknown constants, the same for all sets, and the $\{w_i\}$ are known. We may verify the equality of the proportionality constant σ^2 by evaluating the set of estimators $\{S_i^2\}$ from the mean squares about the set means, $\{\bar{Y}_i\}$, and then using Bartlett's or one of the other tests of Chapter 9 to show that they are all estimators of the same quantity. Assuming that this is the case, we may pool the individual estimators, viz.

$$S_i^2 = \sum_j w_i(Y_{ij} - \bar{Y}_i)^2/(n_i - 1) ,$$

to obtain an unbiased estimator of σ^2, thus

$$S_a^2 = \sum_i\sum_j w_i(Y_{ij} - \bar{Y}_i)^2/(n - r) . \qquad (1)$$

An independent estimator of σ^2 may be obtained from the mean square residual of the set means about the fitted regression line. We shall have r values of $\bar{Y}_i$, each with weight $n_i w_i$, and so our second estimator of σ^2 is

$$S_b^2 = \sum_i n_i w_i(\bar{Y}_i - \sum_{p=0}^{1} B_p x_i^p)^2/(r - 2) . \qquad (2)$$

Under the usual alternative hypothesis that a more complicated polynomial is required, $E(S_b^2) > \sigma^2$ whereas S_a^2 is always unbiased. Thus, a one-sided F test is called for and, working with a significance level α, we reject the null hypothesis that $\mu = \sum_{p=0}^{1} \beta_p x^p$ if $t = S_b^2/S_a^2 \geq F_{r-2,n-r}(1-\alpha)$.

We remark that, for computational purposes, it is desirable to recast equations (1) and (2).

$$s_a^2 = (S_{YY} - S_{\bar{Y}\bar{Y}})/(n - r) ,$$

$$s_b^2 = \{(S_{\bar{Y}\bar{Y}} - S_Y^2)S_w - S_{xY}^2/S_{xx}\}/(r - 2) ,$$

where S_w is the sum of the weights over all the points, that is, $\Sigma n_i w_i$ and $S_{\bar{Y}\bar{Y}} = \Sigma n_i w_i \bar{Y}_i^2$.

6.5. Comparison of several sets of data

A problem which frequently arises is that in which we wish to compare several sets of data. To discuss this problem, we need to alter our subscripting from the previous section, so that $i = 1,2,\ldots r$ designates the set and $j = 1,2,\ldots n_i$ relates to the data within a set. Thus, (w_{ij}, x_{ij}, y_{ij}) represents the experimental data for the jth point of the ith set. We suppose that $V(Y_{ij}) = \sigma_{ij}^2 = \sigma_i^2/w_{ij}$, where σ_i^2 applies throughout the ith set of data.

We could perform tests between the six possible pairs of the hypotheses:

H_a : β_0's all equal, β_1's all equal,

H_b : β_0's all equal, β_1's different,

H_c : β_0's different, β_1's all equal,

H_d : β_0's different, β_1's different.

Here we have selected the main ones, viz. H_b vs H_d (Section 6.5.1), H_c vs H_d (Section 6.5.2), and H_a vs H_d (Section 6.5.3). The remaining tests could be performed similarly or using the methods of Chapter 16. All the tests require that the variances $\{\sigma_i^2\}$ shall be the same, say σ^2, and so firstly we may wish to check this by using Bartlett's or another test of Chapter 9.

6.5.1. Testing the homogeneity of the intercepts

The equation for the error-free line of the ith set may be written

$$\mu = \beta_{0i} + \beta_{1i}x \cdot$$

The two hypotheses to be tested are

$$H_0 : \beta_{01} = \beta_{02} = \ldots \beta_{0r} = \beta_0, \quad \text{say},$$

$$H_1 : \text{not } H_0 \ .$$

We proceed by obtaining one estimator of σ^2 from the scatter of the individual estimators $\{B_{0i}\}$ about their grand mean, and a second estimator from the combined scatter of the individual observations within each set about the regression for that set.

Under H_0, the MVU estimator, B_0, of β_0 is the weighted mean of the individual estimators, thus

$$B_0 = \Sigma w(B_{0i})B_{0i}/\Sigma w(B_{0i}) \ , \tag{1}$$

where $w(B_{0i})$ is the weight associated with B_{0i}, given by

$$\left.\begin{aligned} w(B_{0i}) &= \sigma^2/\sigma^2(B_{0i}) \\ &= \frac{\sum_j w_{ij}\sum_j w_{ij}(x_{ij}-\bar{x}_i)^2}{\sum_j w_{ij}x_{ij}^2} \ . \end{aligned}\right\} \tag{2}$$

We have seen that $B_{0i} \sim N\{\beta_{0i},\sigma^2(B_{0i})\}$, and the B_{0i}'s are clearly independent. Hence, under H_0,

$$\Sigma(B_{0i} - B_0)^2/\sigma^2(B_{0i}) \sim \chi^2_{r-1} \ ,$$

since there are r sets altogether. That is, $\Sigma w(B_{0i})(B_{0i}-B_0)^2/\sigma^2$ is distributed as χ^2_{r-1} so that $\Sigma w(B_{0i})(B_{0i}-B_0)^2/(r-1)$ is the 'between sets' estimator of σ^2, under H_0.

We now obtain an independent estimator of σ^2 from the within-sets scatter of the experimental points about the fitted lines. For the ith set, we know that $\sum_j w_{ij}D_{ij}^2/(n_i-2)$ is an estimator of σ^2. Hence, we may pool these r estimators to obtain a 'within-sets' estimator of σ^2 based on $\Sigma(n_i-2) = n-2r$ degrees of freedom, where $n = \Sigma n_i$ is the total number of data points. This is $\sum_i\sum_j w_{ij} D_{ij}^2/(n-2r)$. Hence, under H_0, the test ratio

$$T = \frac{\Sigma w(B_{0i})(B_{0i} - B_0)^2/(r - 1)}{\sum_i\sum_j w_{ij} D_{ij}^2/(n - 2r)}$$

is distributed as $F_{r-1,n-2r}$. We thus decide whether to reject or accept H_0 at a significance level α according to whether or not $t \geq F_{r-1,n-2r}(1-\alpha)$. If we accept H_0, the best estimator of the intercept is B_0. To find the appropriate estimator of $\sigma^2(B_0)$, we return to equation (1). We obtain

$$\sigma^2(B_0) = \sigma^2/\Sigma w(B_{0i}) \ . \tag{3}$$

Hence

$$S^2(B_0) = S^2/\Sigma w(B_{0i})$$

$$= \frac{\sum_i\sum_j w_{ij} D_{ij}^2/(n - 2r)}{\Sigma w(B_{0i})} \ . \tag{4}$$

We note that B_0 is a linear combination of normally distributed variates $\{B_{0i}\}$, each of which is an unbiased estimator of β_0, (under H_0). Hence B_0 itself is normally distributed and unbiased. We may thus use the standard error of B_0 from equation (4) and b_0 to calculate confidence intervals for β_0.

A few words of caution are necessary when dealing with the case where all the data points in all the sets are of equal weight, that is $\sigma_{ij}^2 = \sigma^2$, all i,j. The above equations do not simplify quite so markedly as is usual because the weighting factors for the individual estimators $\{B_{0i}\}$ will not generally be equal. This is immediately apparent when we examine the simplified form of equation (2), viz.

$$w(B_{0i}) = \frac{n_i \sum_j (x_{ij} - \bar{x}_i)^2}{\sum_j x_{ij}^2} .$$

In general, n_i, $\sum_j (x_{ij} - \bar{x}_i)^2$, and $\sum_j x_{ij}^2$ will vary from one set to another which confirms the above point.

6.5.2. Testing the homogeneity of the slopes

Here we wish to decide whether or not the error-free lines are parallel. The procedure and logic are precisely analogous to that of the previous section, so that it is not necessary to go through all the details. The hypotheses under test are

$$H_0 : \beta_{11} = \beta_{12} = \ldots \beta_{1r} = \beta_1 , \quad \text{say},$$

$$H_1 : \text{not } H_0 .$$

The best estimator of β_1, under H_0, is B_1, the weighted mean of the individual estimators, viz.

$$B_1 = \Sigma w(B_{1i}) B_{1i} / \Sigma w(B_{1i}) , \qquad (1)$$

where

$$\left. \begin{aligned} w(B_{1i}) &= \sigma^2/\sigma^2(B_{1i}) \\ &= \sum_j w_{ij}(x_{ij} - \bar{x}_i)^2 . \end{aligned} \right\} \qquad (2)$$

By analogy with the arguments of the previous section, our test criterion is

$$T = \frac{\Sigma w(B_{1i})(B_{1i} - B_1)^2/(r - 1)}{\sum_i \sum_j w_{ij} D_{ij}^2/(n - 2r)} ,$$

which is distributed as $F_{r-1,n-2r}$ under H_0. If the test leads to the acceptance of H_0, the best estimator of the slope is B_1, calculated by equation (1).

To find the appropriate estimator of $\sigma^2(B_1)$ we proceed as in the previous section and obtain

$$s^2(B_1) = \frac{\sum_{ij}\sum w_{ij}D_{ij}^2/(n - 2r)}{\Sigma w(B_{1i})} . \quad (3)$$

B_1 is normally distributed and unbiased and hence confidence intervals for β_1 may be obtained using b_1 and equation (3).

Again we draw attention to the fact that the expressions do not simplify overmuch when all the points are of the same weight.

6.5.3. Testing for coincidence of the lines

We are here concerned with testing whether or not our r sets of data are associated with the same error-free line or with different lines. Under H_0 we assume that the β_0's are all equal and the β_1's are all equal. Under H_1 we assume that

$$Y_{ij} = \beta_{0i} + \beta_{1i}x_{ij} + \varepsilon_{ij}, \quad i = 1,2,...r, \quad j = 1,2,...n_i ,$$

where $\varepsilon_{ij} \sim N(0,\sigma^2/w_{ij})$. We have the hypotheses

$$H_0 : \beta_{01} = \beta_{02} = ... \beta_{0r} = \beta_0 , \text{ say,}$$

$$\beta_{11} = \beta_{12} = ... \beta_{1r} = \beta_1 , \text{ say,}$$

$$H_1 : \text{not } H_0 .$$

We proceed to test H_0 analogously to the corresponding situation in Chapter 13. We combine all the data together to obtain

$$S_{DD,c} = \sum_{ij}\sum w_{ij}D_{ij,c}^2 ,$$

where

$$D_{ij,c} = Y_{ij} - \sum_p B_p x_{ij}^p .$$

We may recast $S_{DD,c}$ in a form more suitable for computation, thus

$$S_{DD,c} = S'_{YY} - S'^2_{xY}/S'_{xx} ,$$

where the primes as before indicate that the sums are to be taken over the displacements of the suffixed quantities from their means. Thus,

$$S'_{YY} = \sum_{i}\sum_{j} w_{ij}(Y_{ij} - \bar{Y})^2$$

$$= S_{YY} - S_Y^2/S_w ,$$

$$S'_{xY} = \sum_{i}\sum_{j} w_{ij}(x_{ij} - \bar{x})(Y_{ij} - \bar{Y})$$

$$= S_{xY} - S_x S_Y/S_w ,$$

$$S'_{xx} = \sum_{i}\sum_{j} w_{ij}(x_{ij} - \bar{x})^2$$

$$= S_{xx} - S_x^2/S_w ,$$

where $\bar{x}$ and $\bar{Y}$ are the grand means over all the sets, namely,

$$\bar{x} = \sum_{i}\sum_{j} w_{ij}x_{ij}/\sum_{i}\sum_{j} w_{ij} = S_x/S_w ,$$

and

$$\bar{Y} = \sum_{i}\sum_{j} w_{ij}Y_{ij}/\sum_{i}\sum_{j} w_{ij} = S_Y/S_w .$$

We may also calculate a sum of squared residuals for each set; thus for the ith set

$$S_{DD,i} = \sum_{j} w_{ij}D_{ij}^2 ,$$

where

$$D_{ij} = Y_{ij} - \sum_{p} B_{pi}x_{ij}^p .$$

We may sum the r quantities $\{S_{DD,i}\}$ to calculate S_{DD}, given by

$$S_{DD} = \Sigma S_{DD,i}$$

$$= \Sigma S'_{YY,i} - \Sigma(S'^2_{xY,i}/S'_{xx,i}) ,$$

where the primed quantities are defined similarly to those given above; as there, we may express these in simpler terms, thus

$$S'_{YY,i} = \sum_j w_{ij}(Y_{ij} - \bar{Y}_i)^2$$

$$= S_{YY,i} - S^2_{Y,i}/S_{w,i} ,$$

$$S'_{xY,i} = \sum_j w_{ij}(x_{ij} - \bar{x}_i)(Y_{ij} - \bar{Y}_i)$$

$$= S_{xY,i} - S_{x,i}S_{Y,i}/S_{w,i} ,$$

$$S'_{xx,i} = \sum_j w_{ij}(x_{ij} - \bar{x}_i)^2$$

$$= S^2_{xx,i} - S^2_{x,i}/S_{w,i} ,$$

where, as can be seen from the subscripting, the various quantities relate to the ith set.

It can be shown that $(S_{DD,c} - S_{DD})$ and S_{DD} are independently distributed and that the test criterion,

$$T = \frac{(S_{DD,c} - S_{DD})/\{2(r - 1)\}}{S_{DD}/(n - 2r)} ,$$

is distributed as $F_{2(r-1),n-2r}$ under H_0. The numerator and denominator of T are both unbiased estimators of σ^2 under H_0, whilst under H_1 the denominator is still an unbiased estimator, but the numerator has a larger expectation. Accordingly we reject H_0 at the significance level α if $t \geq F_{2(r-1),n-2}(1-\alpha)$.

If the data pass the test, we use B_0 and B_1 calculated from the combined data as our estimators of intercept and slope. We use $S_{DD,c}/(n-2)$ as our estimator of σ^2, treating all our data as if they were a single set, from which we may calculate $S^2(B_0)$ and $S^2(B_1)$ by the usual formulae.

If the data do not pass the test, there is always the possibility that one particular set is markedly different from the rest, and that the remaining r-1 sets are in fact homogeneous. This possibility should be considered before concluding that all the sets are heterogeneous and careful inspection of the intercepts and slopes for the individual sets should be made with this in mind. An analogous discussion is given in Section 10.4 of Chapter 13. In the event that the r sets are not homogeneous, the best estimators of $\{\beta_{0,i}\}$ and $\{\beta_{1,i}\}$ are $\{B_{0,i}\}$ and $\{B_{1,i}\}$ while $S_{DD}/(n-2r)$ again serves as an unbiased estimator of σ^2.

7. Consistency of future experiments

As in Section 11 of Chapter 13, we describe a test which could be used to check the consistency of estimators $B_{0,f}$ and $B_{1,f}$ obtained in some future experiment with data derived from a reasonably large number of previously acquired sets of data. We must assume that the unknown error variances of both the new and the old data are the same.

Our two hypotheses are

$$H_0 : \underline{\beta}_f = \underline{\beta} \quad ,$$

$$H_1 : \underline{\beta}_f \neq \underline{\beta} \quad ,$$

where the elements of the vector $\underline{\beta}_f$ are $\beta_{0,f}$ and $\beta_{1,f}$, parameters estimated by $B_{0,f}$ and $B_{1,f}$; $\underline{\beta}$ stands for the vector of parameters associated with our combined set of previous data. Under H_0, the test criterion,

$$T = (\underline{B}_f - \underline{\beta})'\underline{V}^{-1}(\underline{B}_f - \underline{\beta})/\sigma^2 \quad , \tag{1}$$

is distributed as χ_2^2 , whereas under H_1, the distribution of T is non-central χ_2^2 . Hence a one-sided χ^2 test is appropriate here. $\sigma^2\underline{V}$ is the covariance matrix associated with the vector $\underline{B}_f$.

To carry out the test we replace $\underline{\beta}$ and σ^2 by the estimates $\underline{b}_c$ and s^2 from the consolidated old data. We reject H_0 at the α level of significance if $t \geq \chi_2^2(1-\alpha)$. As before, because of approximations involved, we protect H_0 by using a 1 percent significance level. If the new data pass the test, they should be combined with the old so as to update $\underline{b}_c$ and s^2 for use in subsequent tests of consistency.

We consider it useful to present the arguments in a somewhat different way so as to link Section 2 of Chapter 11 with this and the corresponding section of Chapter 13. When $\underline{B} \sim N(\underline{\beta}, \sigma^2\underline{V})$, we have seen that the quantity M given by

$$M = (\underline{B} - \underline{\beta})'\underline{V}^{-1}(\underline{B} - \underline{\beta})/\sigma^2 \sim \chi_2^2 .$$

Now in the present context, the elements of $(\underline{B} - \underline{\beta})$ and $\sigma^2\underline{V}$ are

$$\begin{pmatrix} B_0 - \beta_0 \\ B_1 - \beta_1 \end{pmatrix} \quad \text{and} \quad \begin{pmatrix} V(B_0) & C(B_0,B_1) \\ C(B_0,B_1) & V(B_1) \end{pmatrix} .$$

Hence

$$M = \left[\frac{(B_0-\beta_0)^2}{V(B_0)} - \frac{2\rho(B_0-\beta_0)(B_1-\beta_1)}{\{V(B_0)V(B_1)\}^{\frac{1}{2}}} + \frac{(B_1-\beta_1)^2}{V(B_1)} \right] / (1 - \rho^2) ,$$

ρ here being $\rho(B_0,B_1) = C(B_0,B_1)/\{V(B_0)V(B_1)\}^{\frac{1}{2}}$. This is a particular application of the general result given in Section 2 of Chapter 11 where we discussed the density function for two jointly normally distributed random variables. The above expression may be written in terms of the weighting factors for the intercept and slope introduced earlier in Sections 6.5.1 and 6.5.2. We deduce that M is given by

$$\frac{w(B_0)(B_0-\beta_0)^2 - 2\rho\{w(B_0)w(B_1)\}^{\frac{1}{2}}(B_0-\beta_0)(B_1-\beta_1) + w(B_1)(B_1-\beta_1)^2}{\sigma^2(1-\rho^2)} .$$

Returning now to our test of consistency, B_0 and B_1 may be replaced by the estimators $B_{0,f}$ and $B_{1,f}$ which we observe in our future experiment. Then, putting

$$\Delta b_0 = b_{0,f} - b_{0,c} ,$$

$$\Delta b_1 = b_{1,f} - b_{1,c} ,$$

and replacing σ^2 by the estimate s^2 obtained from our old set of data, the realization, t, of our test criterion becomes

$$\frac{w(B_{0,f})\Delta^2 b_0 - 2\rho\{w(B_{0,f})w(B_{1,f})\}^{\frac{1}{2}}\Delta b_0 \Delta b_1 + w(B_{1,f})\Delta^2 b_1}{s^2(1-\rho^2)} .$$

We have seen in Section 3.3 that $\rho = -\bar{x}/(\overline{x^2})^{\frac{1}{2}}$ and so t is easily evaluated for a given set of future data.

To see how the test statistic works, we consider a set of data in which $\bar{x} > 0$ so that $\rho < 0$. There are clearly four possibilities:

(a) $\Delta b_0 > 0$, $\Delta b_1 > 0$ (intercept and slope are too 'high')

(b) $\Delta b_0 > 0$, $\Delta b_1 < 0$ (intercept too 'high', slope too 'low')

(c) $\Delta b_0 < 0$, $\Delta b_1 > 0$ (intercept too 'low', slope too 'high')

(d) $\Delta b_0 < 0$, $\Delta b_1 < 0$ (intercept and slope are too 'low')

Now when $\rho < 0$, either of the two events (b) or (c) is much more probable than either event (a) or (d) as may be seen by considering the expression for the joint density, $f(b_0,b_1)$. For events (b) or (c), we see that the second term in the numerator of our test statistic tends to reduce t, which thus accommodates reasonable departures of intercept and slope from the 'true' values $b_{0,c}$ and $b_{1,c}$ when they are in opposite directions. However, the occurrence of Δb_0 and Δb_1 of the same sign, both too 'high' or too 'low' has low probability, and this is reflected in the intolerance of the test to such events.

When $\bar{x} < 0$ and so $\rho > 0$, the usual situation is that high slopes are associated with high intercepts, and so on. Again the test statistic accommodates such events while events of low probability, that is, Δb_0 and Δb_1 of the opposite sign, tend to raise the test statistic to exceed the critical value so that only small departures of $b_{0,f}$ and $b_{1,f}$ from the true values are acceptable if these are in the opposite direction.

Example 1. The e.m.f., E, of a Clark cell varies linearly with temperature T, over a short temperature interval because the entropy change, ΔS, accompanying the cell reaction is a very slowly-varying function of temperature. The relevant equation is

$$zF\left(\frac{\partial E}{\partial T}\right)_p = \Delta S , \tag{1}$$

where z is the number of electrons transferred in the reaction and F is the Faraday constant. Two other thermodynamic parameters of interest are the changes in the Gibbs function and enthalpy, ΔG and ΔH, which accompany the cell reaction

$$Zn(s) + Hg_2SO_4(s) + 7H_2O(l) = ZnSO_4.7H_2O(s) + 2Hg(l) .$$

These quantities are calculated from the equations

$$\Delta G = -zF\,E , \tag{2}$$

$$\Delta H = \Delta G + T\Delta S . \tag{3}$$

In an experiment designed to measure ΔS, $\Delta G_{298.2K}$ and $\Delta H_{298.2K}$ the e.m.f. of a particular Clark cell was measured at a series of randomly selected temperatures between 273 and 313 K. The results are presented in Table 14.1 in terms of the dimensionless variables

$$Y = E/\text{Volt} = E/V , \tag{4}$$

$$x = T/\text{Kelvin} = T/K. \tag{5}$$

Because of the small range covered by Y and the way in which the apparatus functioned, it was reasonable to assume that

$$Y = \alpha_0 + \alpha_1 T_1(x) + \varepsilon , \tag{6}$$

with an error term of constant variance over the range of x.

Table 14.1. The e.m.f. of a Clark cell at a series of temperatures

T/K	E/V
273.2	1.4520
274.5	1.4511
289.7	1.4308
295.2	1.4257
297.2	1.4200
303.0	1.4116
311.3	1.3985

Results from the statistical analysis. From the equations in the main text putting $w_i = 1$, all i, we obtain

$$a_0 = 1.42710 ,$$

$$a_1 = -1.387 \times 10^{-3} ,$$

as our least squares estimates of α_0 and α_1. We also obtain

$$s^2 = 2.91 \times 10^{-6} ,$$

$$s^2(A_0) = 0.415 \times 10^{-6} ,$$

$$s^2(A_1) = 2.42 \times 10^{-9} ,$$

as our estimates of σ^2, $V(A_0)$, and $V(A_1)$, from which we fix the number of significant figures quoted for a_0 and a_1. Each of these estimates has five degrees of freedom.

Estimation of ΔS. From equations (1), (4), (5), and (6),

$$\hat{\Delta S} = zF\, a_1 \text{ V K}^{-1} \qquad (7)$$

$$= -267.656 \text{ J mol}^{-1} \text{ K}^{-1} ,$$

since $z = 2$ and $F = 9.64873 \times 10^4$ C mol^{-1}.

We also have

$$s^2(\hat{\Delta S}) = z^2F^2\, s^2(A_1)\; V^2\, K^{-2}$$

$$= 90.1\ J^2\ mol^{-2}\ K^{-2}\ ,$$

giving $s(\hat{\Delta S}) = 9.49\ J\ mol^{-1}\ K^{-1}$. Thus, our final values for $\hat{\Delta S}$ and its standard error rounded to the appropriate number of significant figures are $-267.7\ J\ mol^{-1}\ K^{-1}$ and $9.5\ J\ mol^{-1}\ K^{-1}$, the latter quantity having 5 degrees of freedom.

Estimation of $\Delta G_{298.2K}$ From equations (2), (4), (5) and (6),

$$\hat{\Delta G}_{298.2K} = -zF\ \hat{\mu}_{298.2K}\ V$$

$$= -zF\ \{a_0 + a_1T_1(298.2)\}\ V\ . \tag{8}$$

Since $T_1(x) = x - \bar{x} = x - 292.013$, $T_1(298.2) = 6.187$. Hence

$$\hat{\Delta G}_{298,2K} = -273.738\ kJ\ mol^{-1}\ .$$

We also have

$$s^2(\hat{\Delta G}_{298.2K}) = z^2F^2\ \{s^2(A_0) + 6.187^2 s^2(A_1)\}\ V^2$$

$$= 1.890 \times 10^{-2}\ kJ^2\ mol^{-2}\ ,$$

so that $s(\hat{\Delta G}_{298.2K})$ is 0.137. Hence, our final values for $\hat{\Delta G}_{298.2K}$ and its standard error rounded to the appropriate number of significant figures are $-273.74\ kJ\ mol^{-1}$ and $0.14\ kJ\ mol^{-1}$, the latter quantity with 5 degrees of freedom.

Estimation of $\Delta H_{298.2K}$ From equations (3), (7) and (8)

$$\frac{\hat{\Delta H}_{298.2K}}{zF\ V} = -\hat{\mu}_{298.2} + 298.2\ a_1 \tag{9}$$

$$= -\ \{a_0 + a_1T_1(298.2)\} + 298.2a_1$$

$$= -a_0 + a_1\bar{x}\ . \tag{10}$$

Hence $\hat{\Delta H}_{298.2K} = -353.553\ kJ\ mol^{-1}$. It follows from equation (10) that

$$s^2(\hat{\Delta H}_{298.2K}) = z^2F^2\{s^2(A_0) + \bar{x}^{-2}s^2(A_1)\} V^2$$
$$= 7.700 \text{ kJ}^2 \text{ mol}^{-2} ,$$

giving $s(\hat{\Delta H}_{298.2K}) = 2.77$ kJ mol^{-1}. Our final values for $\hat{\Delta H}_{298.2K}$ and its standard error are thus -353.6 kJ mol^{-1} and 2.8 kJ mol^{-1}, expressed to the appropriate number of significant figures; as before, our estimate of standard error has 5 degrees of freedom.

We observe that $\hat{\Delta H}$ (or ΔH, for that matter) is independent of temperature which is consistent with the assumption that ΔS is independent of temperature.

A note on the formulation of the model. It will be observed that we have chosen to write our model of the data in the form

$$Y = \alpha_0 + \alpha_1(x - \bar{x}) + \varepsilon ,$$

rather than the more conventional representation

$$Y = \beta_0 + \beta_1 x + \varepsilon .$$

Apart from the mathematical advantages already dealt with in the main text, and illustrated here in the derivation of formulae for variances, there is a good physical reason for the choice, since we do not really want to estimate the value of β_0, the hypothetical error-free e.m.f. at the absolute zero.

Example 2. When a solution is separated from solvent by a membrane which is permeable only to the solvent, any net flow of solvent across the membrane from solvent to solution can be prevented by applying a pressure to the solution in excess of that applied to the solvent. The difference in pressure is called the osmotic pressure, ρ. When the solute is a high molecular weight polymer at a sufficiently low concentration, c, ρ varies with c according to the equation

$$\frac{\rho}{c} = \frac{RT}{\bar{M}_n}(1 + \Gamma_2 c) , \qquad (1)$$

where R is the gas constant, T the absolute temperature, $\bar{M}_n$ the number average molecular weight of the polymer, and Γ_2 the second virial coefficient, which depends on the particular polymer-solvent system and the temperature. Measurements of osmotic pressure are commonly carried out on a series of polymer solutions of differing concentration with the object of estimating $\bar{M}_n$ and Γ_2. This example is concerned with the procedure employed to analyse such data.

In Table 14.2, the osmotic pressures are shown in N m^{-2} for a series of concentrations expressed in kg m^{-3}. For convenience we define

$$\pi = \frac{\rho}{N\ m^{-2}} ,$$

and
$$x = \frac{c}{kg\ m^{-3}} .$$

In terms of π and x, equation (1) becomes

$$\frac{\pi}{x} = \beta_0 + \beta_1 x , \tag{2}$$

where
$$\beta_0 = (RT/\bar{M}_n)\ s^2\ m^{-2} ,$$

and
$$\beta_1 = (RT\Gamma_2/\bar{M}_n)\ kg\ m^{-5}\ s^2 .$$

Of course, we cannot measure π without error, but we observe a random variable P such that

$$P = \pi + \eta ,$$

where η is random error. Thus, an appropriate model for our observations is

$$P/x = \beta_0 + \beta_1 x + \eta/x , \tag{3}$$

which we may rewrite as

$$Y = \beta_0 + \beta_1 x + \varepsilon \tag{4}$$

to correspond with the notation of this chapter. Here $Y = P/x$ and $\varepsilon = \eta/x$ are random variables derived from P and η by division by x.

Table 14.2. Osmotic pressures of a series of polymer solutions of varying concentration and derived quantities

$x = \frac{c}{\text{kg m}^{-3}}$	$p = \frac{\text{observed value of osmotic pressure}}{\text{N m}^{-2}}$				
	RUN 1	RUN 2	RUN 3	RUN 4	RUN 5
0.904	12.26	15.21	13.52	12.68	13.10
3.112	51.98	52.83	53.25	51.98	50.29
3.290	56.21	58.32	55.79	56.21	55.79
5.710	114.11	117.49	115.80	114.11	119.60
7.780	177.92	180.03	176.65	179.61	182.57

$x = \frac{c}{\text{kg m}^{-3}}$	$y = \frac{p}{x}$				
	RUN 1	RUN 2	RUN 3	RUN 4	RUN 5
0.904	13.56	16.83	14.96	14.02	14.49
3.112	16.70	16.98	17.11	16.70	16.16
3.290	17.08	17.73	16.96	17.08	16.96
5.710	19.98	20.58	20.28	19.98	20.95
7.780	22.87	23.14	22.71	23.09	23.47

Notes

(a) These data refer to a toluene solution of a high molecular weight polystyrene at 310.0 K (see also example 15.1 for work on a similar system).

(b) Each osmotic pressure was calculated from the hydrostatic head, h, of toluene necessary to prevent osmosis. The appropriate expression is

$$\pi = hdg,$$

where d is the density of the solvent and g is the intensity of the gravitational field.

If $\eta \sim N(0,\sigma^2)$ then $\varepsilon \sim N(0,\sigma^2/x^2)$, so that $\sigma^2(Y) = \sigma^2/x^2$, decreasing with increasing x. Thus if the precision of our observations is independent of x, the precision of our dependent variable rises as x increases.

The assumption of a constant variance, σ^2, for the error term η may be verified by calculating the estimates of σ^2 from replicate determinations at the various concentrations, as shown in Table 14.3, and performing Bartlett's test. We obtain $s^2 = \Sigma\nu_i s_i^2/\Sigma\nu_i = 2.86$ and t (see Chapter 9, Section 3.2) = 4.70. This latter value is much less than $\chi_4^2(0.95) = 9.49$ and so we accept at the five percent level the hypothesis that $\sigma^2(P)$ is independent of x. The pooled estimate, $s^2 = 2.86$, may be taken as the best available estimate of σ^2. We remark that the same test performed on the $\{s^2(Y)\}$ yields a result which is significant at the five percent level.

Table 14.3. Data for Bartlett's test

x	$s^2(P)$	ν
0.904	1.30	4
3.112	1.29	4
3.290	1.12	4
5.710	5.53	4
7.780	5.06	4

Clearly then we have here a situation which requires the use of weighting factors, different from unity, given by

$$w(Y) = \sigma^2/\sigma^2(Y) = x^2 .$$

It is now a simple matter to test the goodness of fit of the data of Table 14.2 to equation (4) using the procedure of Section 6.4. The quantity S_a^2 of equation (6.4.1) is, in fact, the s^2 calculated in Bartlett's test on the $s^2(P)$. The other quantity S_b^2 is the mean square about the least squares line through the points $\{n_i w_i, x_i, \bar{y}_i\}$. We obtain $s_b^2 = 1.36 < s_a^2$ and so we accept the hypothesis that equation (4) is a suitable model for our data.

Table 14.4 Data used in the test for coincidence of the error-free lines

Set 1			Set 2			Set 3			Set 4		
w_{1j}	x_{1j}	y_{1j}	w_{2j}	x_{2j}	y_{2j}	w_{3j}	x_{3j}	y_{3j}	w_{4j}	x_{4j}	y_{4j}
5.018	2.240	16.23	4.080	2.020	15.90	4.301	2.074	15.69	4.219	2.054	16.46
12.81	3.580	18.42	10.43	3.230	18.06	11.01	3.318	17.70	10.80	3.286	17.49
45.16	6.720	22.01	36.72	6.060	21.20	38.71	6.222	21.67	37.97	6.162	21.67
80.28	8.960	25.75	65.29	8.080	24.58	68.82	8.296	24.81	67.50	8.216	24.28
125.44	11.200	28.83	102.01	10.100	27.70	107.54	10.370	27.75	105.47	10.270	27.49

Set 1	Set 2	Set 3	Set 4
$b_{01} = 12.80$	$b_{02} = 12.53$	$b_{03} = 12.76$	$b_{04} = 12.91$
$b_{11} = 1.431$	$b_{12} = 1.495$	$b_{13} = 1.446$	$b_{14} = 1.411$
$\sum w_{1j} d_{1j}^2 = 12.15$	$\sum w_{2j} d_{2j}^2 = 11.69$	$\sum w_{3j} d_{3j}^2 = 0.7531$	$\sum w_{4j} d_{4j}^2 = 6.192$
$s_1^2 = 4.05$	$s_2^2 = 3.90$	$s_3^2 = 0.251$	$s_4^2 = 2.06$

For the combined data : $b_0 = 12.80$ $\quad \sum_i \sum_j w_{ij} d_{ij,c}^2 = 48.682$

$b_1 = 1.439$ $\quad s^2 = 2.70$

We are now in a position to test a number of sets of data obtained using different sets $\{x_i\}$ to see whether or not they are associated with the same error-free line or with different lines. We use the notation of Section 6.5.3, our data consisting of the triads $\{w_{ij}, x_{ij}, y_{ij}\}$, $i = 1,2,\ldots 4$, $j = 1,2,\ldots 5$, (4 sets, 5 points to each set). These data are given in Table 14.4 together with the estimates b_{0i}, b_{1i}, and s_i^2, and the sums of squares about the regression $\sum_j w_{ij} d_{ij}^2 = S_{DD,i}$. Also given in the table are the corresponding quantities for the regression line through the combined data.

Firstly, we note that by use of Bartlett's test the four independent quantities, $\{s_i^2\}$ may be taken as estimates of the same quantity. The pooled estimate, $\Sigma \nu_i s_i^2 / \Sigma \nu_i$, which is used in this test is identical with what we have called $S_{DD}/(n-2r)$ in the present context, the denominator of our test criterion for coincidence. Its value is 2.57, gratifyingly consistent with our previous estimate of $\sigma^2 = \sigma^2(P)$, calculated from our work on replication and goodness of fit.

Secondly, we require the numerator of our test criterion for coincidence . From the data of Table 14.4, the relevant quantities are

$$S_{DD,c} = \sum_{ij}\sum w_{ij} d_{ij,c}^2 = 48.682, \quad \nu = 18 ;$$

$$S_{DD} = \sum_{ij}\sum w_{ij} d_{ij}^2 = 30.785, \quad \nu = 12 ;$$

$$S_{DD,c} - S_{DD} = 17.897 , \quad \nu = 6 ,$$

so that the numerator in question is 2.98. Hence, our test criterion is $2.98/2.57 = 1.16 < F_{6,18}(0.95) = 2.66$. We thus accept at the five percent significance level the hypothesis that the four error-free lines are coincident.

In this event, the best estimates of β_0, β_1 and σ^2 are those calculated from the combined data, viz.

$$b_0 = 12.80 ,$$

$$b_1 = 1.439 ,$$

$$s^2 = 2.70 ,$$

$$\nu = 18 .$$

The estimates of the variances and covariances of B_0 and B_1 are

$$s^2(B_0) = 4.75 \times 10^{-2} ,$$

$$s^2(B_1) = 5.92 \times 10^{-4} ,$$

$$\hat{C}(B_0,B_1) = -5.14 \times 10^{-3} ,$$

each associated with 18 degrees of freedom.

Incidentally, we could recast equation (3) in the form

$$P = \beta_0 x + \beta_1 x^2 + \eta ,$$

and use the methods described in Chapter 13 to estimate the parameters β_0 and β_1, fitting a least squares quadratic through the combined set of equally weighted (x,P) data and through the origin. In this case we would obtain

$$b_0 = 12.80 ,$$

$$b_1 = 1.439 ,$$

$$\sum_i \sum_j w_{ij} d^2_{ij,c} = 48.604 ,$$

$$s^2 = 2.70 ,$$

which agree extremely well with the estimates derived using the weighted least squares line.

At the beginning of this example, we gave expressions relating $\bar{M}_n$ and Γ_2 to β_0 and β_1. From these expressions, it follows that

$$\hat{\bar{M}}_n = (RT/b_0)\ s^2\ m^{-2}$$

$$= 8.314 \times 310.0/12.80\ \text{J mol}^{-1}\ \text{K}^{-1}\ \text{K s}^2\ \text{m}^{-2}$$

$$= 201.4\ \text{kg mol}^{-1}$$

and

$$\hat{\Gamma}_2 = b_1/b_0\ \text{m}^3\ \text{kg}^{-1}$$

$$= 1.439/12.80\ \text{m}^3\ \text{kg}^{-1}$$

$$= 0.1124\ \text{m}^3\ \text{kg}^{-1} .$$

The corresponding standard errors are given by

$$s(\hat{\bar{M}}_n) = \hat{\bar{M}}_n \, s(B_0)/b_0$$

$$= 3.5 \text{ kg mol}^{-1} ,$$

$$\frac{s(\hat{\Gamma}_2)}{m^3 kg^{-1}} = \left\{ \frac{b_1^2 s^2(B_0)}{b_0^4} + \frac{s^2(B_1)}{b_0^2} - \frac{2b_1 \hat{C}(B_0,B_1)}{b_0^3} \right\}^{\frac{1}{2}}$$

$$= \frac{b_1}{b_0} \left\{ \frac{s^2(B_0)}{b_0^2} + \frac{s^2(B_1)}{b_1^2} - \frac{2\hat{C}(B_0,B_1)}{b_0 b_1} \right\}^{\frac{1}{2}}$$

$$= 3.8 \times 10^{-3} .$$

Since $\bar{M}_n$ and Γ_2 are not normally distributed, we cannot calculate confidence intervals for the corresponding distribution parameters from appropriate values of Student's t and the above standard errors. However, since B_0 is normally distributed, an interval for $\bar{M}_n$ may be obtained by first calculating an interval for β_0 in the usual way and then using the relation $\bar{M}_n = (RT/\beta_0)\ s^2\ m^{-2}$. For example, a 95 percent interval for β_0 is (12.34,13.26) which gives (194.4,208.9) kg mol^{-1} as the corresponding interval for $\bar{M}_n$, an interval which is not quite symmetrical about the estimate of 201.4 kg mol^{-1} obtained previously. Alternatively, following Chapter 5, Section 6, we might appeal to the Central Limit Theorem to assume approximate normality for a function of a set of estimators, though clearly when there are only a few such random variables, as here, the approximation is very crude. Nevertheless, we could give a rough 95 percent confidence interval as the estimated mean ± 2 x standard error. So for $\bar{M}_n$ and Γ_2, our rough confidence intervals are (194.4,208.4) kg mol^{-1} and (0.1048,0.1200) m^3 kg^{-1}. We note that this latter interval for $\bar{M}_n$ agrees well with that calculated from the interval for β_0, due partly to the fact that $t_{18}(0.975) \approx 2$.

Finally, we use the results from our combined data to demonstrate the consistency check procedure described in Section 7. An experiment was performed subsequent to those described earlier, from which we obtained the following estimates

$$b_0 = 14.17 ,$$

$$b_1 = 1.368 ,$$

$$s^2 = 8.28 ,$$

$$n = 5 .$$

(Using a two-sided χ^2 test, this value of s^2 is consistent with our assumed value of $\sigma^2 = 2.70$). The values of $w(B_0)$, $w(B_1)$ and $\rho(B_0,B_1)$ were calculated from the values of $\{w_i,x_i\}$ as

$$w(B_0) = 14.33 ,$$

$$w(B_1) = 1183.8 ,$$

$$\rho(B_0,B_1) = 0.9702 .$$

We obtain

$$t = 52.2 ,$$

which is considerably greater than $\chi_2^2(0.99) = 9.21$. Hence our new data are inconsistent with the old.

Incidentally, this is a case where the formulation of our model in terms of the β's is essential since β_0 is related to a physically important quantity, the number average molecular weight.

CHAPTER 15

THE GENERAL POLYNOMIAL

A natural extension of the problem which we discussed in Chapter 14 is that in which we desire to fit a polynomial through a set of data points $\{w_i, x_i, y_i\}$. As we have repeatedly stressed, this procedure will be the most valuable if there is some good physical reason to suppose that a polynomial is appropriate. However, even without this, a polynomial representation often serves as a useful approximation to the true mean curve, where this is continuous. Such a representation may be useful, for instance, for interpolation purposes or for smoothing the original data prior to further analysis.

As in the case of the general straight line, we shall not discuss the 'unweighted' and 'weighted' cases separately since the equations applicable to the former can be obtained from those appropriate to the latter by simply putting $w_i = 1$, all i.

1. Formal statement of the model

We shall assume that the relationship between the random variable Y and the variable x is of the form

$$Y = \sum_{p=0}^{m} \beta_p x^p + \varepsilon . \tag{1}$$

Here, m represents the degree of the polynomial and the elements of the set $\{\beta_p\}$, p = 0,1,...m, are constants appropriate to the particular situation. As usual, ε represents the error term with the properties

$$E(\varepsilon) = 0 \quad \text{for all } x , \tag{2}$$

$$V(\varepsilon) = \sigma_i^2 \text{ for } x = x_i . \tag{3}$$

Equation (1) differs from its counterpart of Chapter 13 by the inclusion of the leading term β_0. Again, as before, we introduce a set of weights $\{w_i\}$, and a constant σ^2 such that each element of this set is related to the corresponding σ_i^2 by the expression

$$w_i = \sigma^2/\sigma_i^2 . \tag{4}$$

It follows from equations (1), (2), and (3) that

$$E(Y) = \mu = \Sigma\beta_p x^p , \tag{5}$$

$$V(Y_i) = \sigma_i^2 . \tag{6}$$

2. Estimation using the method of least squares

We deduce expressions for the estimators $\{B_p\}$ of $\{\beta_p\}$ by minimising the sum of squares

$$\$ = \Sigma w_i(Y_i - \mu_i)^2 . \tag{1}$$

We obtain a set of m+1 simultaneous equations by equating $\partial\$/\partial\beta_p$ to zero and replacing β_p by B_p $(p = 0,1,\ldots m)$, to give

$$\sum_i w_i(Y_i - \sum_q \beta_q x_i^q) x_i^p = 0, \quad p = 0,1,\ldots m.$$

These equations yield explicit expressions for $\{B_p\}$ and thence for $\hat{\mu}$, viz.

$$\hat{\mu} = \Sigma B_p x^p . \tag{2}$$

An improved notation utilises the quantities

$$\theta_p = \sum_i w_i x_i^p Y_i , \tag{3}$$

$$\phi_{pq} = \sum_i w_i x_i^p x_i^q , \tag{4}$$

precisely as in previous chapters. The m+1 normal equations may then be written

$$\sum_{q=0}^{m} B_q \phi_{pq} = \theta_p , \quad p = 0,1,\ldots m , \tag{5}$$

which in matrix form are

$$\underline{\Phi}\underline{B} = \underline{\theta} . \qquad (6)$$

$\underline{\Phi}$ is an (m+1) x (m+1) symmetrical matrix and $\underline{B}$ and $\underline{\theta}$ are each column vectors containing m+1 elements. The vector of solutions is

$$\underline{B} = \underline{\Phi}^{-1}\underline{\theta} , \qquad (7)$$

where $\underline{\Phi}^{-1}$ stands for the inverse matrix of $\underline{\Phi}$.

3. A look back over the last three chapters

It is worth drawing together the results obtained for the estimators of the various β's defined in the last few chapters. The expressions for μ are:

$$\mu = \beta_1 x , \qquad \text{(Chapter 12)}$$

$$\mu = \sum_{p=1}^{m} \beta_p x^p , \qquad \text{(Chapter 13)}$$

$$\mu = \beta_0 + \beta_1 x , \qquad \text{(Chapter 14)}$$

$$\mu = \sum_{p=0}^{m} \beta_p x^p . \qquad \text{(This chapter)}$$

In each case, the least squares condition leads to a vector of solutions for the B's:

$$\underline{B} = \underline{\Phi}^{-1}\underline{\theta} , \qquad (1)$$

in which the elements of the matrix $\underline{\Phi}$ are given by $\phi_{pq} = \sum_i w_i x_i^{p+q}$, and the elements of the vector $\underline{\theta}$ by $\theta_p = \sum_i w_i x_i^p Y_i$. Here the suffices p and q relate to the pth row and the qth column in the first two cases and to the (p+1)th row and (q+1)th column in the latter two cases. In the first case above, $\underline{\Phi}$ is a 1 x 1 matrix whose single element is $\Sigma w_i x_i^2$; its inverse is, of course, simply $1/\Sigma w_i x_i^2$. Likewise for this particular case, $\underline{\theta}$ consists of the single element $\Sigma w_i x_i Y_i$ so that we recover the expression for B_1 directly from equation (1). For the remaining cases, we have either m x m matrices (Chapter 13), or (m+1) x (m+1) matrices (Chapters 14 and 15); the vector $\underline{\theta}$ contains correspondingly either m or (m+1) elements.

The fact that equation (1) gives the solution to the estimation problem for all four cases means that, if the calculations are performed by computer, only one program needs to be written to find the values of the estimates $\{b_p\}$.

4. Estimation using the method of least squares and polynomials $\{T_p(x)\}$, p = 0,1,...m

We have seen in the previous chapters that considerable simplification may be achieved by recasting the problem in terms of orthogonal polynomials. For the present purpose, we define our polynomials through the equation

$$T_p(x) = \sum_{q=0}^{p} f_{pq} x^q ; \quad f_{pp} = 1, \quad p = 0,1,\ldots m . \quad (1)$$

The first two elements of the set $\{T_p(x)\}$ have already been shown in Chapter 14 to be

$$T_0(x) = 1 ,$$

$$T_1(x) = x - \bar{x} ,$$

where $\bar{x} = \Sigma w_i x_i / \Sigma w_i$. The remaining elements can be calculated as in Appendix 2.

4.1. Reformulation of our model

We proceed as before to write

$$Y = \sum_{p=0}^{m} \alpha_p T_p(x) + \varepsilon , \quad (1)$$

where the error term has the same properties as in Section 1.

It is an easy matter to show that the estimators $\{A_p\}$ of $\{\alpha_p\}$ are given by

$$A_p = \frac{\sum_i w_i Y_i T_p(x_i)}{\sum_i w_i T_p^2(x_i)} , \qquad p = 0,1,\ldots m , \tag{2}$$

so that our estimator, $\hat{\mu}$, of μ may be obtained from

$$\hat{\mu} = \Sigma A_p T_p(x) . \tag{3}$$

We note that

$$A_0 = \frac{\Sigma w_i Y_i}{\Sigma w_i} = \bar{Y} , \tag{4}$$

as in Chapter 14.

4.2. Expectations of the estimators

Since the A's and B's are linear combinations of the Y's, it follows that

$$E(A_p) = \alpha_p ,$$

$$E(B_p) = \beta_p .$$

Hence $E(\hat{\mu}) = \mu.$

All the estimators are thus unbiased.

4.3. The covariance matrix of the $\{A_p\}$ and the variance of $\hat{\mu}$

The arguments of Chapter 13, Section 5.2 show that

$$C(A_p, A_q) = \frac{\sigma^2 \sum_i w_i T_p(x_i) T_q(x_i)}{\left\{\sum_i w_i T_p^2(x_i)\right\}\left\{\sum_i w_i T_q^2(x_i)\right\}} . \tag{1}$$

Hence, by the orthogonality property,

$$C(A_p, A_q) = 0 \text{ when } p \neq q. \tag{2}$$

Furthermore

$$V(A_p) = \frac{\sigma^2}{\sum_i w_i T_p^2(x_i)} , \qquad (3)$$

a result which could have been obtained directly from equation (4.1.2). Since

$$\hat{\mu} = \Sigma A_p T_p(x) ,$$

$$V(\hat{\mu}) = \Sigma T_p^2(x) V(A_p) , \qquad (4)$$

because the covariances $C(A_p, A_q)$, $p \neq q$, are zero. Substituting for $V(A_p)$, we obtain

$$V(\hat{\mu}) = \sigma^2 \sum_p \frac{T_p^2(x)}{\sum_i w_i T_p^2(x_i)} . \qquad (5)$$

4.4. The covariance matrix of $\{B_p\}$

First we require a relationship between the A's and the B's. By comparing coefficients of x^r in the identities

$$\hat{\mu} = \Sigma B_p x^p = \sum_p A_p \left(\sum_{q=0}^{p} f_{pq} x^q \right) ,$$

we obtain

$$B_r = \sum_{p=r}^{m} f_{pr} A_p . \qquad (1)$$

An analogous procedure connects the α's and β's, thus

$$\beta_r = \sum_{p=r}^{m} f_{pr} \alpha_p . \qquad (2)$$

Now, to obtain $C(B_r, B_s)$, we proceed as follows

$$C(B_r, B_s) = E\{(B_r - \beta_r)(B_s - \beta_s)\}$$

$$= E\left[\{ \sum_{p=r}^{m} f_{pr}(A_p - \alpha_p) \} \{ \sum_{p=s}^{m} f_{ps}(A_s - \alpha_s) \} \right]$$

$$= \sum_{p=\max(r,s)}^{m} f_{pr} f_{ps} V(A_p) . \qquad (3)$$

$$= \sigma^2 \sum_{p=\max(r,s)}^{m} \{f_{pr} f_{ps} / \sum_i w_i T_p^2(x_i)\} . \quad (4)$$

In particular, with s = r,

$$V(B_r) = \sigma^2 \sum_{p=r}^{m} \{f_{pr}^2 / \sum_i w_i T_p^2(x_i)\} . \quad (5)$$

4.5. <u>The weighted sum of squared residuals and derived variance estimators</u>

We have seen in Chapter 13, Section 5.4, that the weighted sum of squared residuals,

$$\Sigma w_i D_i^2 = \Sigma w_i (Y_i - \hat{\mu}_i)^2 ,$$

may be written

$$\Sigma w_i D_i^2 = \Sigma w_i Y_i^2 - \Sigma w_i \hat{\mu}_i^2 . \quad (1)$$

If we now substitute for $\hat{\mu}_i$ using the representation involving the polynomials $\{T_p(x_i)\}$, we obtain

$$\Sigma w_i D_i^2 = \Sigma w_i Y_i^2 - \sum_{p=0}^{m} A_p^2 \sum_i w_i T_p^2(x_i) \quad (2)$$

$$= \Sigma w_i (Y_i - \bar{Y})^2 - \sum_{p=1}^{m} A_p \sum_i w_i T_p^2(x_i) , \quad (3)$$

since $A_0 = \bar{Y}$ and $T_0(x_i) = 1$, all i. From this last expression we observe that the total sum of squares accounted for by the regression is

$$\sum_{p=1}^{m} A_p^2 \sum_i w_i T_p^2(x_i) .$$

We shall return to this result when testing for the most appropriate order for the polynomial.

The expectation of $\Sigma w_i D_i^2$ follows from equation (1), thus

$$E(\Sigma w_i D_i^2) = \Sigma w_i E(Y_i^2) - \Sigma w_i E(\hat{\mu}_i^2)$$

$$= \Sigma w_i V(Y_i) - \Sigma w_i V(\hat{\mu}_i)$$

$$= n\sigma^2 - \sigma^2 \sum_i w_i \sum_p \frac{T_p^2(x_i)}{\sum_i w_i T_p^2(x_i)}$$

$$= \{n - (m+1)\}\sigma^2 , \qquad (4)$$

since there are m+1 terms in the sum. Hence

$$s^2 = \frac{\Sigma w_i D_i^2}{n-m-1}$$

is an unbiased estimator of σ^2. As usual, we observe that each of the m+1 parameters defining the error-free curve reduces $n\sigma^2$ by σ^2.

Now that we have an estimator of σ^2, we may write down expressions for the estimators of variance and covariance. These are

$$s^2(A_p) = \frac{s^2}{\sum_i w_i T_p^2(x_i)} , \qquad (5)$$

$$s^2(B_p) = s^2 \sum_{r=p}^{m} \frac{f_{rp}^2}{\sum_i w_i T_r^2(x_i)} , \qquad (6)$$

$$s^2(\hat{\mu}) = s^2 \sum_p \frac{T_p^2(x)}{\sum_i w_i T_p^2(x_i)} , \qquad (7)$$

$$\hat{C}(B_p, B_q) = s^2 \sum_{r=\max(p,q)}^{m} \frac{f_{rp} f_{rq}}{\sum_i w_i T_r^2(x_i)} . \qquad (8)$$

4.6. Altering the order of the polynomial

We have seen in Chapter 13 how easy it is to deal with a change in the order of the fitted polynomial if the representation in terms of $\{T_p(x)\}$ is used. We assume that we have completed the estimation of the $\{\alpha_p\}$, $\{\mu_i\}$, σ^2 etc., for a polynomial of order m. Then for an increase of 1 in the order we have

$$A'_p = A_p, \quad p = 0,1,\ldots m , \tag{1}$$

$$A'_{m+1} = \frac{\Sigma w_i Y_i T_{m+1}(x_i)}{\Sigma w_i T^2_{m+1}(x_i)} , \tag{2}$$

$$\hat{\mu}' = \hat{\mu} + A'_{m+1} T_{m+1}(x) , \tag{3}$$

$$S'^2 = \frac{\Sigma w_i D'^2_i}{n-(m+2)} , \tag{4}$$

$$\Sigma w_i D'^2_i = \Sigma w_i D^2_i - A'^2_{m+1} \Sigma w_i T^2_{m+1}(x_i) , \tag{5}$$

where the primes denote that the quantities refer to a fitted polynomial of order m+1.

5. Sampling distributions assuming the normality of the data

When the $\{Y_i\}$ are independently, normally distributed, we have

$$\left.\begin{array}{l} A_p \sim N\{\alpha_p, \sigma^2(A_p)\} \\ \\ B_p \sim N\{\beta_p, \sigma^2(B_p)\} \end{array}\right| \quad p = 0,1,\ldots m ,$$

since each estimator is a linear combination of the $\{Y_i\}$. Also, these estimators belong to the class of maximum likelihood estimators. Furthermore

$$\hat{\mu} \sim N\{\mu, \sigma^2(\hat{\mu})\}$$

and $\quad \Sigma w_i D^2_i/\sigma^2 = (n-m-1)S^2/\sigma^2 \sim \chi^2_{n-m-1}$

independently of the distributions of the $\{A_p\}$, $\{B_p\}$, and $\hat{\mu}$. Finally, the $\{A_p\}$ are independent.

6. Calculation of confidence intervals

The results quoted in Section 5 enable confidence intervals for the parameters $\{\alpha_p\}$, $\{\beta_p\}$ and μ to be easily calculated. For example,

$$b_p \pm t_{n-m-1}(1-\alpha/2)s(B_p)$$

represents an interval for β_p with coefficient $1-\alpha$.

7. Tests of a number of hypotheses

Here, we follow the practice of previous chapters, testing hypotheses chiefly relating to $\{\beta_p\}$, but not specifically to $\{\alpha_p\}$ because such a set is 'tied' to a particular set of x_i. However, we may still use $\{A_p\}$ to calculate some of the quantities required for our tests, e.g. $\Sigma w_i D_i^2$, since this set will often provide the easiest method of computation.

7.1. Tests on $\{\beta_p\}$ and μ

The usual tests of a null hypothetical value, say $\beta_{p,0}$ or μ_0, may be performed. We have, for example, the hypotheses

$$H_0 : \beta_p = \beta_{p,0} \quad \left| \begin{array}{l} H_1 : \beta_p > \beta_{p,0} , \\ H_1 : \beta_p < \beta_{p,0} , \quad \text{or} \\ H_1 : \beta_p \neq \beta_{p,0} , \end{array} \right.$$

when, under H_0, $(B_p-\beta_{p,0})/S(B_p) \sim t_{n-m-1}$.

When our two hypotheses are

$$H_0 : \beta_p = 0 ,$$

$$H_1 : \beta_p \neq 0 ,$$

the test amounts to deciding whether or not to include the term in x^p in the power series representation of μ. This test could be used, therefore, to decide upon the best order to adopt for the fitted polynomial, truncating the series at order m, say, where m is the largest integer for which the estimate of β_m differs significantly from zero. A better procedure, however, is given in Section 7.4.

7.2. Test of goodness of fit of the model

We have referred to this problem several times already in previous chapters so that only the briefest mention is now necessary. In this test, we compare an estimator of σ^2 obtained by replication with an independent estimator of σ^2 obtained from the residuals.

The details are precisely the same as those given in Chapter 14, Section 6.4 except that the divisor for $\Sigma w_i D_i^2$ is now n-m-1.

7.3. Comparison of several sets of data

Here we wish to test whether or not r sets of data points, the ith of which is $\{w_{ij}, x_{ij}, Y_{ij}\}$, $j = 1,2,...n_i$, all have the same underlying model

$$Y = \Sigma\beta_p x^p + \varepsilon , \tag{1}$$

where ε is random error, with zero expectation. We assume that each of the r sets has a model of the form of equation (1) with coefficients $\{\beta_{pi}\}$ so that we have to test between the two hypotheses

$$H_0 : \beta_{p1} = \beta_{p2} = ... \beta_{pr} = \beta_p, \text{ say, all } p ,$$

$$H_1 : \text{not } H_0 .$$

In general, the variance of the error term will depend on both the set to which Y belongs and the value of x at which Y is measured, thus $V(\varepsilon_{ij}) = \sigma_{ij}^2 = \sigma_i^2/w_{ij}$. However, for this analysis, we shall assume that $\sigma_i^2 = \sigma^2$, an unknown constant, for all i, an assumption which may be tested, if necessary,by the methods of Chapter 9.

If H_O is accepted, we may ignore the distinction of the data according to set. We may thus combine all the data together to calculate a set of estimators $\{B_p\}$ and a sum of squares

$$S_{DD,c} = \sum_i\sum_j w_{ij} D_{ij,c}^2 ,$$

where $$D_{ij,c} = Y_{ij} - \sum_p B_p x_{ij}^p .$$

To calculate $S_{DD,c}$ it will be most convenient to use

$$S_{DD,c} = \sum_i\sum_j w_{ij}(Y_{ij}-\bar{Y})^2 - \sum_p A_p^2 \sum_i\sum_j w_{ij} T_p^2(x_{ij}) ,$$

where $\bar{Y} = \sum_i\sum_j w_{ij}Y_{ij}/\sum_i\sum_j w_{ij}$, and $\{A_p\}$ and $\{T_p(x)\}$ are computed using the combined set of all the data. We may also compute a sum of squares, S_{DD}, from the scatter of the individual points within sets. We put

$$S_{DD} = \Sigma S_{DD,i} ,$$

where $$S_{DD,i} = \sum_j w_{ij} D_{ij}^2 ,$$

and $$D_{ij} = Y_{ij} - \sum_p B_{pi} x_{ij}^p ,$$

$\{B_{pi}\}$ being the estimators of $\{\beta_{pi}\}$ calculated from the ith set of data. Again for ease of computation, we may use

$$S_{DD} = \sum_i\sum_j w_{ij}(Y_{ij}-\bar{Y}_i)^2 - \sum_i\{\sum_p A_{pi}^2 \sum_j w_{ij} T_{pi}^2(x_{ij})\} ,$$

where $\bar{Y}_i = \sum_j w_{ij}Y_{ij}/\sum_j w_{ij}$, and A_{Oi}, $A_{1i}, \ldots A_{mi}$ and the corresponding T_{pi}'s relate to the ith set of data. $S_{DD,c}$ is associated with $\Sigma n_i-(m+1) = n-m-1$ degrees of freedom, there being just m+1 estimated parameters, while the part of S_{DD} due to the variation of the ith set of data about its fitted polynomial has n_i-m-1 degrees of freedom, so that S_{DD} is associated with $n-r(m+1)$ degrees of freedom. Further, it can be shown that $S_{DD,c} - S_{DD}$ is distributed independently of S_{DD}.

As in Chapter 13, Section 10.4, we test H_0 against H_1 using the criterion

$$T = \frac{(S_{DD,c} - S_{DD})/\{(r-1)(m+1)\}}{S_{DD}/\{n-r(m+1)\}} .$$

Under H_0, the numerator and denominator of T are independent, unbiased estimators of the same quantity, σ^2, and so T is distributed as $F_{(r-1)(m+1),n-r(m+1)}$. Under H_1, the denominator still provides an unbiased estimator of σ^2, but the numerator has an expectation greater than σ^2. Hence, we use a one-sided test and reject H_0 at significance level α if $t \geq F_{(r-1)(m+1),n-r(m+1)}(1-\alpha)$.

If we accept H_0, we use $\{B_p\}$ as the estimators of $\{\beta_p\}$, while if we reject H_0 we use $\{B_{pi}\}$ as the estimators of $\{\beta_{pi}\}$. In both cases, $S_{DD}/\{n-r(m+1)\}$ provides an estimator of σ^2, but as stated previously $S_{DD,c}/(n-m-1)$ is more appropriate when H_0 is accepted.

If a large number of data for a particular experimental situation have been accumulated and accepted as homogeneous, we may use the resulting estimates of $\{\beta_p\}$ and σ^2 as a basis for an approximate test of consistency of future data following Section 11 of Chapter 13.

7.4. How to decide upon the most appropriate order of polynomial to be used

We discussed this problem at some length in Chapter 13, Section 10.3. It will be recalled that we compare systematically a series of models, using the weighted sum of squared residuals for each model as our guide to the final degree of polynomial. We summarise the procedure below.

Model	Relation of μ to x	$\Sigma w_i D_i^2$
0	$\mu = \alpha_0 T_0(x)$	$S_{DD,0}$
1	$\mu = \alpha_0 T_0(x) + \alpha_1 T_1(x)$	$S_{DD,1}$
2	$\mu = \alpha_0 T_0(x) + \alpha_1 T_1(x) + \alpha_2 T_2(x)$	$S_{DD,2}$

etc.

The index number of the model corresponds to the highest power of x which appears in the relation of μ to x. We also choose to discuss the problem in terms of the T_p's because of (a) the simplicity of the explanation and (b) the convenience of the calculation of $S_{DD,1}$, $S_{DD,2}$, etc.[see equation (4.5.3)].

Our first two hypotheses are

$$H_0 : \mu = \alpha_0 T_0(x) \ ,$$

$$H_1 : \mu = \alpha_0 T_0(x) + \alpha_1 T_1(x) \ .$$

We now form the following quantities

$$A_1 = \frac{\Sigma w_i Y_i T_1(x_i)}{\Sigma w_i T_1^2(x_i)} \ ,$$

$$S_{DD,0} = \Sigma w_i (Y_i - \bar{Y})^2 \ ,$$

$$S_{DD,1} = S_{DD,0} - A_1^2 \Sigma w_i T_1^2(x_i) \ .$$

These equations lead to the following conclusions

	Under H_0	Under H_1
$E(S_{DD,0})$	$(n-1)\sigma^2$	$(n-1)\sigma^2 + \alpha_1^2 \Sigma w_i T_1^2(x_i)$
$E(S_{DD,1})$	$(n-2)\sigma^2$	$(n-2)\sigma^2$
$E(S_{DD,0} - S_{DD,1})$	σ^2	$\sigma^2 + \alpha_1^2 \Sigma w_i T_1^2(x_i)$

Now it may be shown that $(S_{DD,0} - S_{DD,1})\sigma^2$ and $S_{DD,1}/\sigma^2$ are independently distributed as χ_1^2 and χ_{n-2}^2 respectively under H_0. Hence

$$T = \frac{S_{DD,0} - S_{DD,1}}{S_{DD,1}/(n-2)}$$

is distributed as $F_{1,n-2}$ under H_0. In contrast, the numerator of T will contain a contribution from the term $\alpha_1^2 \Sigma w_i T_1^2(x_i)$ if H_1 applies and so we require a one-sided test to decide between H_O and H_1.

If we decide to accept H_1 against the null hypothesis H_O, we may investigate the new hypotheses

$$H_O : \mu = \alpha_O T_O(x) + \alpha_1 T_1(x) \quad ,$$

$$H_1 : \mu = \alpha_O T_O(x) + \alpha_1 T_1(x) + \alpha_2 T_2(x) \quad .$$

For the test of H_O against H_1, we form the following quantities

$$A_2 = \frac{\Sigma w_i Y_i T_2(x_i)}{\Sigma w_i T_2^2(x_i)} \quad ,$$

$$S_{DD,2} = S_{DD,1} - A_2^2 \Sigma w_i T_2^2(x_i) \quad ,$$

from which we may prove the results tabulated below

	Under H_O	Under H_1
$E(S_{DD,1})$	$(n-2)\sigma^2$	$(n-2)\sigma^2 + \alpha_2^2 \Sigma w_i T_2^2(x_i)$
$E(S_{DD,2})$	$(n-3)\sigma^2$	$(n-3)\sigma^2$
$E(S_{DD,1} - S_{DD,2})$	σ^2	$\sigma^2 + \alpha_2^2 \Sigma w_i T_2^2(x_i)$

Then

$$T = \frac{S_{DD,1} - S_{DD,2}}{S_{DD,2}/(n-3)} \sim F_{1,n-3} \quad , \quad \text{under } H_O \quad ,$$

which provides, as before, a one-sided test to accept/reject H_O.

We may continue in this way to find the most appropriate order of polynomial by which to represent our data.

Example 1. In Chapter 14, we discussed some osmotic pressure data obtained at a series of mass concentrations, c, of a high molecular weight polymer. There we used the relation

$$\frac{\rho}{c} = \frac{RT}{\bar{M}_n}\left(1 + \Gamma_2 c\right) \tag{1}$$

to represent the variation of osmotic pressure, ρ, with c. This equation is an approximation to the quite general relation

$$\frac{\rho}{c} = \frac{RT}{\bar{M}_n}\left(1 + \Gamma_2 c + \Gamma_3 c^2 + \ldots\right), \tag{2}$$

where Γ_2, Γ_3 etc. are termed second, third, etc., virial coefficients.

Here, we illustrate how to decide where to truncate the polynomial in c for a particular system. We then go on to estimate the various quantities of theoretical interest, including the ratio $g = \Gamma_3/\Gamma_2^2$.

Following Chapter 14, we rewrite equation (2) in the form

$$\pi/x = \beta_0 + \beta_1 x + \beta_2 x^2 + \ldots \tag{3}$$

where $\pi = \rho/(\mathrm{N\ m^{-2}})$, $x = c/(\mathrm{kg\ m^{-3}})$ and β_0, β_1, β_2 etc. are simply and obviously related to RT, $\bar{M}_n$, Γ_2, and so on. Then, as previously, if

$$Y = \pi/x + \varepsilon ,$$

and the measured osmotic pressures P are independently normally distributed with constant variance σ^2, then $Y \sim N(\pi/x, \sigma^2/x^2)$ showing that $w(Y) = x^2$.

Table 15.1 gives the values of the raw experimental data (heights, h, of toluene equivalent to the osmotic pressures), the derived osmotic pressures p $\mathrm{N\ m^{-2}}$, and $y = p/x$ for a series of x values.

In passing, we note that the heights h could be read on the particular instrument used here to ± 0.005 cm. This determines the number of significant figures which may be quoted for the p's, and consequently for the y's. In the last two entries for y, the decision whether to quote one or two figures after the decimal point was made on the basis that

Table 15.1. The 'raw' data and related quantities

$x = \frac{c}{\text{kg m}^{-3}}$	$\frac{h}{\text{cm toluene}}$	p	y
1.80	0.17	14	8.0
4.36	0.60	51	11.6
6.68	1.13	96	14.3
10.26	2.42	205	19.9
10.98	2.78	235	21.4
17.06	6.34	536	31.4
18.78	7.63	645	34.3
20.32	9.08	768	37.8
24.38	13.30	1124	46.1
25.18	14.34	1212	48.1

Note: The heights h were converted to osmotic pressures p using 0.8616 g cm^{-3} for the density of toluene and 9.81 m s^{-2} as the intensity of the gravitational field.

---

little information would be lost on truncation to one decimal place because of the large coefficient of variation of y. Indeed, one might make a generalization and say that, for most work, only three significant figures are required for the varying part of a set of numbers - thus, for example, 12.134, 12.369, 12.928 etc. would probably contain all the significant information about the source of variation in the context of the particular experiment generating such values. Confirmation that such a procedure is sound is easily obtained (if a computer program is available) by simply rerunning the calculation using more figures after the decimal place and comparing the two sets of estimates and their standard errors. This is perhaps the best way to decide difficult cases prior to publication. While on this problem, we deliberately avoided its discussion in the corresponding situation in Chapter 14, simply quoting all pressures and y's to two decimal places, and then at the end of our calculations quoting our estimates in accordance with our usual rule.

Table 15.2 summarises the results of our computations of the various quantities required for deciding the order of polynomial to use. We see that a quadratic fit is appropriate for these data, a result not unexpected in view of the wider concentration range and higher molecular weight polymer used here compared with the analogous situation of Chapter 14.

Table 15.2. Data for deciding the order of the polynomial (compare Table 13.5)

m	df	$S_{DD,m}$	$s^2 = \frac{S_{DD,m}}{df}$	R_m	$T = \frac{R_m}{s^2}$	$F_{1,df}(0.95)$
0	9	228000	25333	-	-	-
1	8	1475.7	184.46	226524.3	1228.0	5.32
2	7	89.948	12.850	1385.752	107.8	5.59
3	6	72.957	12.160	16.991	1.4	5.99
4	5	72.472	14.494	0.485	<1	6.61
5	4	71.306	17.827	1.166	<1	7.71

Our estimates of β_0, β_1, and β_2 together with their standard errors are

$$b_0 = 6.82 \ , \qquad s(B_0) = 0.72 \ ,$$

$$b_1 = 1.022 \ , \qquad s(B_1) = 0.084 \ ,$$

$$b_2 = 0.0243 \ , \qquad s(B_2) = 0.0023 \ ,$$

each standard error being associated with 7 degrees of freedom. Other statistics which we shall require are

$$s^2 = 12.85 \ ,$$

$$\hat{C}(B_0,B_1) = -5.85 \times 10^{-2} \ ,$$

$$\hat{C}(B_0,B_2) = 1.54 \times 10^{-3} \ ,$$

$$\hat{C}(B_1,B_2) = -1.94 \times 10^{-4} \ .$$

Since the temperature of the experiment was 310.2 K, these data show that the estimated number average molecular weight of the polymer is 378 kg mol^{-1} with a standard error of 40 kg mol^{-1} having 7 degrees of freedom.

It is instructive to examine the contributions of the three terms of equation (3) to $\hat{\mu}$. These are shown in Table 15.3. We see that the third term, b_2x^2, is considerably less than the estimated standard deviation of Y for $x < 4.36$. Hence, the effect of this term is likely to be 'lost' in the random error of measurement and the linear approximation

$$\pi/x = \beta_0 + \beta_1 x$$

is adequate at concentrations less than 4.36 kg m^{-3} for this particular polymer/solvent/temperature combination - hence the comment about 'low' c in the example of Chapter 14.

Table 15.3 Contributions to $\hat{\mu}$, etc.

			Contributions to $\hat{\mu}$		
x	y	$s(Y) = \frac{s}{x}$	b_0	b_1x	b_2x^2
1.80	8.0	2.0	6.82	1.84	0.08
4.36	11.6	0.82	6.82	4.46	0.46
6.68	14.3	0.54	6.82	6.83	1.08
10.26	19.9	0.35	6.82	10.49	2.56
10.98	21.4	0.33	6.82	11.22	2.93
17.06	31.4	0.21	6.82	17.44	7.07
18.78	34.3	0.19	6.82	19.19	8.57
20.32	37.8	0.18	6.82	20.77	10.03
24.38	46.1	0.15	6.82	24.92	14.44
25.18	48.1	0.14	6.82	25.73	15.41

Finally, we estimate the quantities Γ_2, Γ_3 and $g = \Gamma_3/\Gamma_2^2$, together with their standard errors. We shall simply quote the appropriate expressions, leaving their derivations to our readers.

(i)
$$\frac{\hat{\Gamma}_2}{m^3\ kg^{-1}} = b_1/b_0$$

$$= 0.14985\ ,$$

$$\frac{s(\hat{\Gamma}_2)}{m^3\ kg^{-1}} = \frac{b_1}{b_0}\left\{\frac{s^2(B_0)}{b_0^2} + \frac{s^2(B_1)}{b_1^2} - \frac{2\hat{C}(B_0,B_1)}{b_0 b_1}\right\}^{\frac{1}{2}}$$

$$= 0.028\ .$$

Hence, our final value for $\hat{\Gamma}_2$ is 0.150 $m^3\ kg^{-1}$ with standard error 0.028 $m^3\ kg^{-1}$ based on 7 degrees of freedom (c. of v. = 19%).

(ii)
$$\frac{\hat{\Gamma}_3}{m^6\ kg^{-2}} = b_2/b_0$$

$$= 3.5630 \times 10^{-3}\ ,$$

$$\frac{s(\hat{\Gamma}_3)}{m^6\ kg^{-2}} = \frac{b_2}{b_0}\left\{\frac{s^2(B_0)}{b_0^2} + \frac{s^2(B_2)}{b_2^2} - \frac{2\hat{C}(B_0,B_2)}{b_0 b_2}\right\}^{\frac{1}{2}}$$

$$= 1.4 \times 10^{-4}\ .$$

Hence, our final value for $\hat{\Gamma}_3$ is $3.56 \times 10^{-3}\ m^6\ kg^{-2}$ with standard error $0.14 \times 10^{-3}\ m^6\ kg^{-2}$ based on 7 degrees of freedom (c. of v. = 3.9%).

(iii)
$$\hat{g} = b_0 b_2/b_1^2$$

$$= 0.15867,$$

Our expression for $s(\hat{g})$ includes the three estimates of covariance: $\hat{C}(B_0,B_1)$, $\hat{C}(B_1,B_2)$, $\hat{C}(B_0,B_2)$, thus,

$$s(\hat{g}) = \frac{b_0 b_2}{b_1^2}\left\{\frac{s^2(B_0)}{b_0^2} + \frac{4s^2(B_1)}{b_1^2} + \frac{s^2(B_2)}{b_2^2} + \frac{2\hat{C}(B_0,B_2)}{b_0 b_2} - \frac{4\hat{C}(B_0,B_1)}{b_0 b_1} - \frac{4\hat{C}(B_1,B_2)}{b_1 b_2}\right\}^{\frac{1}{2}}$$

$$= 0.057 \ .$$

Hence, our final value for $\hat{g}$ is 0.159 with standard error 0.057 based on 7 degrees of freedom (c. of v. = 36%).

Example 2. This example relates to a situation which does not lend itself to the use of weights, although the variances are different for the different Y's. If V(Y) is a known function f, say, of $\mu = \mu(x)$, we may write

$$w(Y) = \sigma^2/V(Y) = \sigma^2/f(\mu) \ .$$

However, although it is sometimes done, it is not correct to approximate $f(\mu)$ by f(Y) here, even when the 'error' in Y is relatively small (that is, the coefficient of variation of Y is small), since we then obtain a weight which is a random variable of non-negligible variance. Indeed if Y and μ are interchangeable in the weight function, they must be interchangeable anywhere else. The appropriate procedure is to transform Y so that the new variable, Z, has an approximately constant variance. See Chapter 9, Section 4.

Of course E(Z) will be a different function of x from μ, often of a less convenient kind to work with for least squares estimation, but many continuous functions can be reasonably approximated by a polynomial of not too large an order.

We now suppose for our example that approximately $V(Y) \propto \mu^2$ (a frequently-occurring situation), that is, the coefficient of variation of Y is approximately constant. We also assume that this coefficient is small. We see in Chapter 9, Section 4 that the log transformation is appropriate, so we put

$$Z = \ln Y \ .$$

We assume that E(Z) is (approximately, at least) a polynomial in x, say

$$E(Z) = \beta_0 + \beta_1 x + \beta_2 x^2 + \ldots \tag{2}$$

We do not know what degree of polynomial is appropriate. Our data are now $\{x_i, z_i\}$, $i = 1,2,\ldots n$, and we proceed to decide what degree to use for future work by applying the method of Section 7.4, which, incidentally, is robust to reasonable departures from normality.

CHAPTER 16

A BRIEF LOOK AT MULTIPLE REGRESSION

In this chapter we look into the testing of the general linear model, for which an observed Y value equals a known linear combination of a set of unknown parameters plus random error. The coefficients of the parameters may be powers of a known variable x (as in the previous four chapters), or any functions of x, or unrelated quantities. The linearity relates to the dependence of E(Y) on the unknown parameters, not to the dependence on x.

This chapter is more theoretical and advanced than the previous ones, and includes the theory of most of Chapters 12 to 15 as particular cases. It should be helpful for those readers who wish to go more deeply into matters.

1. The general linear model - hypotheses and least squares estimation

Using matrix notation, we assume

$$\text{Model 1:} \quad \underline{Y} = \underline{X}\underline{\beta} + \underline{\varepsilon} , \tag{1}$$

where $\underline{Y}$ and $\underline{\varepsilon}$ are nx1 vectors, $\underline{\beta}$ is an mx1 unknown vector and $\underline{X}$ an nxm known matrix of rank m < n.

[In the case where the rank of $\underline{X}$ is $m_0 < m$, we can let $\underline{X}_a$ denote a set of m_0 linearly independent columns of $\underline{X}$ and put $\underline{X} = \underline{X}_a\underline{X}_b$, where the m_0 rows of $\underline{X}_b$ are linearly independent. If we now put $\underline{\beta}_a = \underline{X}_b\underline{\beta}$, we work with the model

$$Y = \underline{X}_a\underline{\beta}_a + \varepsilon .$$

(Incidentally, it can be shown that only linear functions of $\underline{\beta}_a$ can be estimated without bias by a linear function of Y).]

The covariance matrix of $\underline{Y}$, or of $\underline{\varepsilon}$, is $\sigma^2\underline{V}$, where $\underline{V}$ is a known, symmetric, nonsingular nxn matrix, and σ^2 is unknown. We assume

$$E(\underline{\varepsilon}) = \underline{O} .$$

The first column of $\underline{X}$ consists of the n given values of a variable x_1, which are often all 1's, the second column of the given values of a variable x_2, etc.

For any positive nxn matrix $\underline{V}$, say, we can easily show that there exists an nxn nonsingular matrix $\underline{U}$ such that $\underline{U}\underline{U}' = \underline{V}$. We may achieve this by putting $\underline{V} = \underline{M}\underline{\Lambda}\underline{M}'$, and $\underline{U} = \underline{M}\underline{\Lambda}^{\frac{1}{2}}$, where $\underline{\Lambda}$ is the diagonal matrix of eigenvalues of $\underline{V}$, the elements of $\underline{\Lambda}^{\frac{1}{2}}$ are the square roots of those of $\underline{\Lambda}$ and $\underline{M}'\underline{M} = \underline{I}_n$. Here our $\underline{V}$ is the covariance matrix of $\underline{Y}$, $V(\underline{Y})$, and we define

$$\underline{Y}_O = \underline{U}^{-1}\underline{Y} ,$$

$$\underline{\varepsilon}_O = \underline{U}^{-1}\underline{\varepsilon} ,$$

$$\underline{X}_O = \underline{U}^{-1}\underline{X} .$$

Then our model becomes

$$\text{Model 1'}: \quad \underline{Y}_O = \underline{X}_O\underline{\beta} + \underline{\varepsilon}_O , \qquad (2)$$

where

$$V(\underline{Y}_O) = V(\underline{\varepsilon}_O) = \sigma^2\underline{I}_n ,$$

$$E(\underline{\varepsilon}_O) = \underline{O} ,$$

$$E(\underline{Y}_O) = \underline{X}_O\underline{\beta} .$$

Our original model (1) is more general than simply using weighted data points (the weights corresponding to the diagonal elements of $\underline{V}^{-1}$), in that the Y values may also have known correlations between them. Model (1) includes the models of Chapters 12 and 13 (using $x_p = x^p$), and Chapters 14 and 15 (using $x_p = x^{p-1}$), as particular cases. In model (1') we have now transformed the situation to the ordinary unweighted model.

By the method of least squares we estimate $\underline{\beta}$ by $\underline{B}$ so as to minimise

$$\$ = (\underline{Y}_O - \underline{X}_O\underline{\beta})'(\underline{Y}_O - \underline{X}_O\underline{\beta}) \; . \tag{3}$$

It may easily be shown that we take $\underline{B} = \underline{\Phi}^{-1}\underline{\theta}$ (c.f. Chapter 15, Section 3), where

$$\underline{\Phi} = \underline{X}_O'\underline{X}_O = \underline{X}'\underline{V}^{-1}\underline{X} \; , \tag{4}$$

$$\underline{\theta} = \underline{X}_O'\underline{Y}_O = \underline{X}'\underline{V}^{-1}\underline{Y} \; . \tag{5}$$

The sum of squared residuals about the regression is

$$\begin{aligned} S_{DD,O} &= (\underline{Y}_O - \underline{X}_O\underline{\Phi}^{-1}\underline{\theta})'(\underline{Y}_O - \underline{X}_O\underline{\Phi}^{-1}\underline{\theta}) \\ &= \underline{Y}_O'\underline{Y} - \underline{Y}_O'\underline{X}\underline{\Phi}^{-1}\underline{X}_O'\underline{Y}_O \\ &= \underline{Y}_O'\underline{Y}_O - \underline{Y}_O'\underline{X}_O(\underline{X}_O'\underline{X}_O)^{-1}\underline{X}_O'\underline{Y}_O & (6) \\ &= \underline{Y}'\underline{V}^{-1}\underline{Y} - \underline{Y}'\underline{V}^{-1}\underline{X}(\underline{X}'\underline{V}^{-1}\underline{X})^{-1}\underline{X}'\underline{V}^{-1}\underline{Y} \; . & (7) \end{aligned}$$

Here we recognize $\underline{Y}'\underline{V}^{-1}\underline{Y}$ as the total sum of squares and $\underline{Y}'\underline{V}^{-1}\underline{X}(\underline{X}'\underline{V}^{-1}\underline{X})^{-1}\underline{X}'\underline{V}^{-1}\underline{Y}$ as the sum of squares due to the regression.

We assume model (1') [or model (1)] applies under H_O, and suppose that, under H_1,

$$\text{Model 2':} \quad \underline{Y}_O = \underline{X}_1\underline{\beta}_1 + \underline{\varepsilon}_1 \; , \tag{8}$$

where the orders of $\underline{X}_1$, $\underline{\beta}_1$ and $\underline{\varepsilon}_1$ are nxk, kx1 and nx1, $k > m$, the first m columns of $\underline{X}_1$ are the same as those of $\underline{X}_O$, and $\text{rank}(\underline{X}_1) = k < n$. We have $E(\underline{\varepsilon}_1) = \underline{O}$ and $V(\underline{\varepsilon}_1) = \sigma^2\underline{I}_n$. The sum of squared residuals under this model, $S_{DD,1}$ can be defined similarly to $S_{DD,O}$, replacing X_O by X_1 in equation (6). [We could equivalently write

$$\text{Model 2:} \quad \underline{Y} = \underline{X}_2\underline{\beta}_2 + \underline{\varepsilon}_2 \; , \tag{9}$$

where the order of $\underline{X}_2$ is nxk, its first m columns being those of $\underline{X}$, and $V(\underline{\varepsilon}_2) = \sigma^2\underline{V}$].

2. Some related distribution theory

Suppose $\underline{W}$ is any vector random variable distributed as $N(\underline{\eta}, \underline{I}_n)$ and $\underline{H}$ is any nxn matrix of rank r which is idempotent, that is, $\underline{H}^2 = \underline{H}$. Note that for an idempotent matrix, rank = trace. It can easily be shown that $\underline{W}'\underline{H}\underline{W}$ is distributed as non-central χ^2 with r degrees of freedom, and

$$E(\underline{W}'\underline{H}\underline{W}) = r + \underline{\eta}'\underline{H}\underline{\eta} .$$

We may show this by writing $H = \underline{L}\underline{\Lambda}\underline{L}'$, where $\underline{\Lambda}$ is now the nxn eigenvalue matrix of $\underline{H}$ with r diagonal elements equal to 1 and all other elements equal to zero (since the eigenvalues satisfy $\lambda_i^2 = \lambda_i$), and $\underline{L}\underline{L}' = \underline{I}_n$. Then $\underline{W}'\underline{H}\underline{W}$ is the sum of squares of the elements of $\underline{\Lambda}_1\underline{L}'\underline{W}$, where $\underline{\Lambda}_1$ is comprised of the r rows of $\underline{\Lambda}$ containing a 1. The vector $\underline{\Lambda}_1\underline{L}'\underline{W} \sim N(\underline{\Lambda}_1\underline{L}'\underline{\eta}, \underline{I}_r)$. Hence $\underline{W}'\underline{H}\underline{W}$ is the sum of squares of r independent normal variates each with unit variance. Such a quantity is distributed as non-central chi-squared with r degrees of freedom and non-centrality parameter equal to the sum of squared means, which in this case is $\underline{\eta}'\underline{L}\underline{\Lambda}_1'\underline{\Lambda}_1\underline{L}'\underline{\eta} = \underline{\eta}'\underline{L}\underline{\Lambda}\underline{L}'\underline{\eta} = \underline{\eta}'\underline{H}\underline{\eta}$.

When $\underline{H}\underline{\eta} = \underline{O}$, this parameter is zero and $\underline{W}'\underline{H}\underline{W} \sim \chi_r^2$, the ordinary chi-squared with r degrees of freedom. For example, when $\underline{Z} \sim N(\underline{O}, \underline{I}_n)$ and $\underline{H} = \underline{I}_n$, then $\underline{Z}'\underline{Z} \sim \chi_n^2$, as stated in Chapter 13, Section 11.

It can further be shown that, for any symmetric nxn matrices $\underline{L}$ and $\underline{M}$, $\underline{W}'\underline{L}\underline{W}$ (or $\underline{L}\underline{W}$) is independent of $\underline{W}'\underline{M}\underline{W}$ (or $\underline{M}\underline{W}$) if and only if $\underline{L}\underline{M} = \underline{O}$ (or equivalently, $\underline{M}\underline{L} = \underline{O}$). For example, taking the linear forms $\underline{L}\underline{W}$ and $\underline{M}\underline{W}$, their covariance is $C(\underline{L}\underline{W}, \underline{M}\underline{W}) = \underline{L}\underline{I}_n\underline{M}' = \underline{L}\underline{M} = \underline{M}\underline{L}$. Hence if and only if $\underline{L}\underline{M}$ (or $\underline{M}\underline{L}$) $= \underline{O}$, $C(\underline{L}\underline{W}, \underline{M}\underline{W}) = \underline{O}$ which for normal $\underline{W}$ is equivalent to independence. The proof for the quadratic forms is related. In outline, we may factorize $\underline{L}$ into $\underline{K}\underline{K}'$ and $\underline{M}$ into $\underline{J}\underline{J}'$, so that $\underline{W}'\underline{L}\underline{W}$ is a function of $\underline{K}'\underline{W}$ and $\underline{W}'\underline{M}\underline{W}$ is a function of $\underline{J}'\underline{W}$. The quadratic forms are independent if and only if the corresponding linear forms are.

We now apply the above general results to obtain the distributions of $S_{DD,1}$ and $S_{DD,0} - S_{DD,1}$ of Section 1. Consider

$\underline{H} = \underline{C} - \underline{D}$, $\underline{K} = \underline{I}_n - \underline{C}$, where $\underline{C} = \underline{X}_1(\underline{X}_1'\underline{X}_1)^{-1}\underline{X}_1'$, $\underline{D} = \underline{X}_0(\underline{X}_0'\underline{X}_0)^{-1}\underline{X}_0'$, $\underline{X}_0$ and $\underline{X}_1$ being as in Section 1. Clearly $\underline{C}^2 = \underline{C}$, $\underline{D}^2 = \underline{D}$ and $\underline{K}^2 = \underline{K}$.

Now if we partition $\underline{X}_1$ as $(\underline{X}_0|\underline{X}_3)$, then

$$\underline{C}\underline{X}_0 = (\underline{X}_0|\underline{X}_3)\left(\begin{array}{c|c} \underline{X}_0'\underline{X}_0 & \underline{X}_0'\underline{X}_3 \\ \hline \underline{X}_3'\underline{X}_0 & \underline{X}_3'\underline{X}_3 \end{array}\right)^{-1}\left(\begin{array}{c} \underline{X}_0' \\ \hline \underline{X}_3' \end{array}\right)\underline{X}_0$$

$$= (\underline{X}_0|\underline{X}_3)\left(\begin{array}{c|c} \underline{X}_0'\underline{X}_0 & \underline{X}_0'\underline{X}_3 \\ \hline \underline{X}_3'\underline{X}_0 & \underline{X}_3'\underline{X}_3 \end{array}\right)^{-1}\left(\begin{array}{c} \underline{X}_0'\underline{X}_0 \\ \hline \underline{X}_3'\underline{X}_0 \end{array}\right)$$

$$= \text{first m cols. of } (\underline{X}_0|\underline{X}_3)\left(\begin{array}{c|c} \underline{X}_0'\underline{X}_0 & \underline{X}_0'\underline{X}_3 \\ \hline \underline{X}_3'\underline{X}_0 & \underline{X}_3'\underline{X}_3 \end{array}\right)^{-1}\left(\begin{array}{c|c} \underline{X}_0'\underline{X}_0 & \underline{X}_0'\underline{X}_3 \\ \hline \underline{X}_3'\underline{X}_0 & \underline{X}_3'\underline{X}_3 \end{array}\right)$$

$$= \text{first m cols. of } (\underline{X}_0|\underline{X}_3)\left(\begin{array}{c|c} I_m & 0 \\ \hline 0 & I_{k-m} \end{array}\right)$$

$$= \underline{X}_0 .$$

Hence, $\underline{C}\underline{D} = \underline{X}_0(\underline{X}_0'\underline{X}_0)^{-1}\underline{X}_0' = \underline{D}$, (and similarly $\underline{D}\underline{C} = \underline{D}$), so that

$$\underline{H}^2 = \underline{H} \text{ and } \underline{K}\underline{H} = \underline{O} .$$

Further, $\text{rank}(\underline{C}) = k$ and $\text{rank}(\underline{D}) = m$, so that $\text{rank}(\underline{K}) = n-k$ and $\text{rank}(\underline{H}) = k-m$, since the ranks equal the traces. Also $\underline{K}\underline{X}_0 = \underline{O}$ and $\underline{K}\underline{X}_1 = \underline{O}$. Hence

$$\underline{K}\underline{H} = \underline{O} .$$

Consequently $\underline{Y}_0'\underline{K}\underline{Y}_0$ and $\underline{Y}_0'\underline{H}\underline{Y}_0$ are independent. Also under H_0 and H_1

$$E(\underline{Y}_0'\underline{K}\underline{Y}_0) = \sigma^2(n - k) ,$$

and under H_0

$$E(\underline{Y}_0'\underline{H}\underline{Y}_0) = \sigma^2(k - m) ,$$

while under H_1

$$E(\underline{Y}_0'\underline{H}\underline{Y}_0) = \sigma^2(k-m) + \underline{\beta}_1'\underline{X}_1\underline{H}\underline{X}_1\underline{\beta}_1 .$$

Hence $\underline{Y}_0'\underline{H}\underline{Y}_0/\sigma^2$ and $\underline{Y}_0'\underline{K}\underline{Y}_0/\sigma^2$ are independently distributed as chi-squared with k-m and n-k degrees of freedom, while under H_1 the independence still applies and the distribution of $\underline{Y}_0'\underline{K}\underline{Y}_0$ remains the same, but $\underline{Y}_0'\underline{H}\underline{Y}_0$ is distributed as non-central χ^2_{k-m} with non-centrality parameter $\underline{\beta}_1'\underline{X}_1'\underline{H}\underline{X}_1\underline{\beta}_1$.

3. The test

Equation (1.7) gave us the sum of squared residuals under H_0 as

$$S_{DD,0} = \underline{Y}'\underline{V}^{-1}\underline{Y} - \underline{Y}'\underline{V}^{-1}\underline{X}(\underline{X}'\underline{V}^{-1}\underline{X})^{-1}\underline{X}'\underline{V}^{-1}\underline{Y} , \tag{1}$$

which is equal to $\underline{Y}_0'(\underline{I}_n-\underline{D})\underline{Y}_0$. Analogously this sum under H_1 becomes

$$S_{DD,1} = \underline{Y}'\underline{V}^{-1}\underline{Y} - \underline{Y}'\underline{V}^{-1}\underline{X}_2(\underline{X}_2'\underline{V}^{-1}\underline{X}_2)^{-1}\underline{X}_2\underline{V}^{-1}\underline{Y} , \tag{2}$$

which is equal to $\underline{Y}_0'(\underline{I}_n-\underline{C})\underline{Y}_0 = \underline{Y}_0'\underline{K}\underline{Y}_0$.

To test H_0 against H_1, assuming the normality of $\underline{Y}$, we use the test criterion

$$T = \frac{(S_{DD,0}-S_{DD,1})/(k-m)}{S_{DD,1}/(n-k)} , \tag{3}$$

which is distributed as $F_{k-m,n-k}$ under H_0, but tends to be larger than this variate under H_1. Then the expectation of the numerator is greater than σ^2, while that of the denominator remains at σ^2. We reject H_0 at the α level of significance if t exceeds $F_{k-m,n-k}(1-\alpha)$. We mention that nearly all the tests in Chapters 12, 13, 14 and 15, are particular cases of this test.

4. Ordinary weighting

When $\underline{V}$ is diagonal with ith diagonal element w_i^{-1}, so that $\underline{V}^{-1}$ is also diagonal with ith diagonal element w_i, we have situations similar to the previous chapters. In this case we have, referring to equation (3.1),

$$\underline{Y}'\underline{V}^{-1}\underline{Y} = \Sigma w_i Y_i^2 \quad = S_{YY} ,$$

the jth element of $\quad \underline{X}'\underline{V}^{-1}\underline{Y} = \sum_i w_i x_{ji} Y_i \quad = S_{x_j Y}$, and

the (j,k) element of $\underline{X}'\underline{V}^{-1}\underline{X} = \sum_i w_i x_{ji} x_{ki} = S_{x_j x_k}$.

Thus we have all the quantities required to calculate $S_{DD,O}$. Analogous expressions apply for obtaining $S_{DD,1}$.

4.1. Ordinary weighting and $\underline{x}_1 = \underline{1}$, a column of 1's

In this case the first of the normal equations is

$$\Sigma w_i Y_i = B_O \Sigma w_i + B_1 \Sigma w_i x_{2i} + B_2 \Sigma w_i x_{3i} + \ldots + B_{m-1} \Sigma w_i x_{mi} ,$$

or

$$S_Y = B_O S_w + B_1 S_{x_2} + B_2 S_{x_3} + \ldots + B_{m-1} S_{x_m} . \qquad (1)$$

The jth equation, j = 2,3,...m, is

$$\sum_i w_i x_{ji} Y_i = B_O S_{x_j} + B_1 S_{x_j x_2} + B_2 S_{x_j x_3} + \ldots + B_{m-1} S_{x_j x_m} . \quad (2)$$

We multiply equation (1) by S_{x_j}/S_w and subtract from (2), so cancelling the term in B_O, to give us, for j = 2,3,...m,

$$S'_{x_j Y} = B_1 S'_{x_j x_2} + B_2 S'_{x_j x_3} + \ldots + B_{m-1} S'_{x_j x_m} , \qquad (3)$$

where

$$S'_{x_j Y} = S_{x_j Y} - S_{x_j} S_Y / S_w$$

$$= \sum_i w_i (x_{ji} - \bar{x}_j)(Y_i - \bar{Y}) ,$$

$$S'_{x_j x_k} = S_{x_j x_k} - S_{x_j} S_{x_k} / S_w$$

$$= \sum_i w_i (x_{ji} - \bar{x}_j)(x_{ki} - \bar{x}_k), \quad k = 2,3,\ldots m ,$$

as can easily be shown. We now see that we have reduced the set of m equations in m unknowns given by equation (2) to the set of m-1 equations in m-1 unknowns given by equation (3), plus the single equation (1) for one unknown, B_O. This splitting of the normal equations reduces the amount of computation involved in calculating the B's. We see that equation (3) is of the same form as equation (2), simply with S′'s replacing S's and with B_O omitted.

In the particular case when $x_j = x^{j-1}$, j = 1,2,...m, we are dealing with a general polynomial. If we are fitting a polynomial through the origin, we simply put $B_O = 0$ in equation (2), and solve the set of m-1 equations in m-1 unknowns. The corresponding set of m-1 equations in m-1 unknowns in equation (3) relates to the general polynomial of order m-1 not through the origin. The calculations and all derived formulae are just the same, except that we use S's in the first case and S′'s in the second. Note that

$$S'_{x_j x_i} = \sum_i w_i (x_i^j - \overline{x^j})(x_i^k - \overline{x^k})$$

$$\neq \sum_i w_i (x_i^j - \bar{x}^j)(x_i^k - \bar{x}^k) ,$$

where

$$\overline{x^j} = \sum_i w_i x_i^j / \Sigma w_i ,$$

$$\bar{x}^j = (\Sigma w_i x_i / \Sigma w_i)^j ,$$

that is, the sums of squares and cross-products are about the means of the powers of x, not about the powers of the mean of x.

APPENDIX 1

DRAWING A RANDOM SAMPLE USING A TABLE OF RANDOM NUMBERS

Tables of random numbers are tables of the integers 0,1,...9 occurring in random order, having been generated by some suitable randomizing device, which has been tested in various ways to confirm the independence of the series of integers. They are usually tabulated in rows and columns and grouped in sets of convenient size. Such tables are readily available in many different collections of statistical tables.

If we wish to draw a random sample from a batch of size N, it is convenient to number the N items 1,2,...N. Now N will be comprised of m digits, where $10^{m-1} \le N < 10^m$. We put a = the integer part of $10^m/N$. For instance, if N is 29, a is 3. We now associate m-digit numbers (including left zeros) 1,2,...a with item 1, the numbers a+1, a+2,...2a with item 2, and so on. If $aN = 10^m$, we associate the numbers a(N-1)+1, a(N-1)+2,...aN-1 and 0 with item N. We then go to a table of random numbers (using different parts of the table on different occasions), read our first random number, say b, and see which item it corresponds to, which is the one we draw first - provided $b \le aN$, otherwise we discard it. Actually b will correspond to

$$\text{item number} = \begin{cases} b/a, \text{ if this is integer,} \\ \text{or else integer part of } b/a + 1. \end{cases} \quad (1)$$

If we were sampling with replacement, we would just repeat this process until we had our full sample. However, a random sample from a batch of size N refers to sampling <u>without</u> replacement, in which case every time we sample an item that has already been sampled, we discard it, and continue until we have our full sample.

<u>Example</u>. If we wish to draw a random sample of size 4 from a batch of size 37, we number the items of the batch 1,2,...37. Here m = 2 and a = 2, so we associate two 2-digit numbers with each item, using numbers

01 to 74.

Consecutive pairs of digits from a table of random numbers are as follows: 64, 77, 86, 37, 27, 18, 47. By equation (1) these correspond to items 32, -, -, 19, 14, 9, the dashes corresponding to discarded random numbers.

Incidentally, the reason why we have assigned a integers to each item instead of just one is to prevent us having to discard too many random numbers read from the table.

Of course, if a computer is being used, it is much more convenient to generate pseudo-random numbers by a simple subroutine. The multiplicative congruential method may be used, whereby we start with any number x (non-integer, of several digits). Using certain appropriate numbers c_1 and c_2, we put

$$y = \text{remainder when } c_1 x \text{ is divided by } c_2, \text{ and}$$

$$z = y/c_2 \, .$$

The number z is then our random number. We use y as our next x value to generate the next random number, and so on. These random numbers are very nearly independent and rectangularly distributed between 0 and 1. Also consecutive digits in such a number are virtually independent integers between 0 and 9 inclusive. One pair of suitable values for c_1 and c_2 is 32768 and 16775723 respectively (Downham and Roberts, 1967).

APPENDIX 2

ORTHOGONAL POLYNOMIALS IN x

Here, we deal with the computation of orthogonal polynomials in x for use in Chapters 13 and 15. We shall devote most of our attention to the set $\{T_p(x)\}$, $p = 1,2,...m$, which we use to treat the polynomial through the origin or some other fixed point. The computation of the extended set $\{T_p(x)\}$, $p = 0,1,2,...m$, for use with the more general situation is similar in essentials, as we shall see.

1. Computation of $\{T_p(x)\}$, $p = 1,2,...m$

It will be recalled that these polynomials are defined by

$$T_p(x) = \sum_{q=1}^{p} c_{pq} x^q \ , \quad c_{pp} = 1 \ , \tag{1}$$

and satisfy the condition

$$\sum_i w_i T_p(x_i) T_q(x_i) = 0 \ , \ p \neq q \ . \tag{2}$$

We observe that c_{pq} is the coefficient of x^q in the expansion of $T_p(x)$ in terms of x, and that the orthogonality condition, equation (2), applies only at the data points.

The first three polynomials in x are thus

$$T_1(x) = x \ ,$$

$$T_2(x) = c_{21}x + x^2 \ ,$$

$$T_3(x) = c_{31}x + c_{32}x^2 + x^3 \ .$$

Clearly, the higher polynomials may be expressed as a linear combination of polynomials of lower degree. For example, the second and third polynomials in x_i can be shown to be

$$T_2(x_i) = c_{21}T_1(x_i) + x_i^2 \ , \tag{3}$$

$$T_3(x_i) = (c_{31}-c_{32}c_{21})T_1(x_i) + c_{32}T_2(x_i) + x_i^3 \quad . \qquad (4)$$

If now we multiply each side of equation (3) by $w_iT_1(x_i)$ and then sum the result for all values of i, we obtain

$$\sum_i w_iT_1(x_i)T_2(x_i) = c_{21} \sum_i w_iT_1^2(x_i) + \sum_i w_iT_1(x_i)x_i^2 \quad .$$

But the left-hand side of this equation is zero because of the orthogonality condition. Hence

$$c_{21} = - \frac{\Sigma w_iT_1(x_i)x_i^2}{\Sigma w_iT_1^2(x_i)} \quad . \qquad (5)$$

Since $T_1(x_i) = x_i$,

$$c_{21} = - \frac{\Sigma w_ix_i^3}{\Sigma w_ix_i^2} \quad . \qquad (6)$$

Equation (6) enables us to calculate $\{T_2(x_i)\}$. Once these are known, a similar procedure to that employed above gives expressions for the coefficients c_{31} and c_{32} which appear in the equation defining $T_3(x)$. We obtain

$$c_{32} = - \frac{\Sigma w_iT_2(x_i)x_i^3}{\Sigma w_iT_2^2(x_i)} \quad , \qquad (7)$$

$$c_{31} = c_{32}c_{21} - \frac{\Sigma w_iT_1(x_i)x_i^3}{\Sigma w_iT_1^2(x_i)} \quad . \qquad (8)$$

Obviously, when these coefficients are evaluated, $T_3(x_i)$ can be calculated.

It will be clear that this process can be repeated successively through the higher degree polynomials and the coefficients c_{pq} calculated one by one. However, this procedure is more tedious than is really necessary. The reason is that a polynomial $T_p(x_i)$ does not have to be expressed as a linear combination of all the lower degree polynomials, $T_{p-1}(x_i)$, $T_{p-2}(x_i), ..T_1(x_i)$. Only two of these are required as we shall show.

1.1. Recurrence relation for the orthogonal polynomials $\{T_p(x_i)\}$ and the coefficients c_{pq}

Consider a particular polynomial $T_p(x_i)$ in which the highest power of x_i is x_i^p. Obviously the product, $x_i T_p(x_i)$, will have x_i^{p+1} as its highest power and so can be written as a linear combination of the polynomials $T_1(x_i)$ to $T_{p+1}(x_i)$, thus

$$x_i T_p(x_i) = g_{p1}T_1(x_i) + g_{p2}T_2(x_i) + \ldots$$

$$g_{p,p-1}T_{p-1}(x_i) + g_{pp}T_p(x_i) + T_{p+1}(x_i) \quad . \tag{1}$$

g_{pq} is the coefficient of $T_q(x_i)$ in the expansion of $x_i T_p(x_i)$; since the coefficient of x_i^{p+1} in the product $x_i T_p(x_i)$ is one, it follows that $g_{p,p+1}$ is also one and this is the reason why this latter symbol does not occur explicitly in equation (1). By an analogous procedure to that employed previously, we may evaluate $\{g_{pq}\}$. First, we multiply equation (1) by $w_i T_q(x_i)$ and then we sum over i for q = 1,2,...p to obtain

$$\sum_i w_i T_q(x_i) x_i T_p(x_i) = g_{pq} \sum_i w_i T_q^2(x_i) \quad . \tag{2}$$

This can be written as

$$g_{pq} \sum_i w_i T_q^2(x_i) = \sum_i w_i T_p(x_i) \{x_i T_q(x_i)\} \quad .$$

We may replace the quantity $x_i T_q(x_i)$ by the appropriate version of equation (1) to obtain

$$g_{pq} \sum_i w_i T_q^2(x_i) = \sum_i w_i T_p(x_i) \{g_{q1}T_1(x_i) + \ldots$$

$$g_{qq}T_q(x_i) + T_{q+1}(x_i)\} \quad . \tag{3}$$

There are p versions of equation (3) as q runs from 1 to p. Let us take the first p-2 of these. The highest degree of polynomial inside the chain brackets will be T_{p-1} and this will not appear until the final equation of this group is reached. In each of these cases, therefore, the right-hand side of equation (3) is zero, from the orthogonality condition. Hence, we obtain the important result

$$g_{pq} = 0 , \quad q \le p-2 . \tag{4}$$

The (p-1)th version of equation (3), that is the version with q = p-1, yields an expression which we shall use ultimately for calculating $g_{p,p-1}$, namely

$$g_{p,p-1} = \frac{\sum_i w_i T_p^2(x_i)}{\sum_i w_i T_{p-1}^2(x_i)} . \tag{5}$$

The final version of equation (3) with q = p gives no further information.

Equation (4) enables equation (1) to be dramatically simplified. We obtain

$$x_i T_p(x_i) = g_{p,p-1} T_{p-1}(x_i) + g_{pp} T_p(x_i) + T_{p+1}(x_i) . \tag{6}$$

A simple rearrangement of this result gives us an equation by which we may calculate the (p+1)th polynomial given those of degree p and p-1, namely

$$T_{p+1}(x_i) = x_i T_p(x_i) - g_{pp} T_p(x_i) - g_{p,p-1} T_{p-1}(x_i) . \tag{7}$$

Equation (7) may be used to obtain an expression for g_{pp}. Applying the orthogonality condition, we obtain

$$g_{pp} = \frac{\sum_i w_i x_i T_p^2(x_i)}{\sum_i w_i T_p^2(x_i)} . \tag{8}$$

We may also use equation (7) to obtain a recurrence relation for the coefficients $\{c_{pq}\}$. We expand each of the polynomials $T_{p+1}(x_i)$, $T_p(x_i)$, and $T_{p-1}(x_i)$ in terms of x_i and then compare coefficients of x_i^{q+1} to obtain

$$c_{p+1,q+1} = c_{pq} - g_{pp}c_{p,q+1} - g_{p,p-1}c_{p-1,q+1} . \quad (9)$$

In using this equation, we should recall that

$$\left.\begin{aligned} c_{p0} &= 0 , \\ c_{pp} &= 1 , \\ c_{pq} &= 0 , \; q > p \end{aligned}\right\} \quad \text{all } p .$$

1.2 Summary of the computational procedure

It will be useful to give a step-by-step description of the computation of the orthogonal polynomials, $T_p(x_i)$.

(a) Put $T_1(x_i) = x_i$.

(b) Calculate c_{21} using equation (1.5), viz.

$$c_{21} = -\frac{\Sigma w_i x_i^3}{\Sigma w_i x_i^2} .$$

(c) Calculate each of the $T_2(x_i)$ using

$$T_2(x_i) = c_{21}x_i + x_i^2 .$$

(d) Calculate g_{21} and g_{22} using equations (1.1.5) and (1.1.8), viz.

$$g_{21} = \frac{\Sigma w_i T_2^2(x_i)}{\Sigma w_i T_1^2(x_i)} ,$$

$$g_{22} = \frac{\Sigma w_i x_i T_2^2(x_i)}{\Sigma w_i T_2^2(x_i)} .$$

(e) Calculate each of the $T_3(x_i)$ using the recurrence relation, equation (1.1.7), viz.

$$T_3(x_i) = x_i T_2(x_i) - g_{22} T_2(x_i) - g_{21} T_1(x_i) .$$

(f) Calculate g_{32} and g_{33} using equations (1.1.5) and (1.1.8).

(g) Calculate each of the $T_4(x_i)$ using the recurrence relation, equation (1.1.7).

(h) Continue in this way to build up the whole set of orthogonal polynomials.

It will be obvious that an exactly analogous procedure can be followed to obtain the set of coefficients $\{c_{pq}\}$.

1.3 A numerical example

Tables A2.1 and A2.2 give the values of the coefficients $\{c_{pq}\}$ and $\{g_{pq}\}$ which are used in the computation of $T_p(x_i)$ for x_i = 0.5 (0.5) 5.0 and $w_i = 1$, all i. Also shown in the tables are the values of $\{T_p(x_i)\}$. In Table A2.1, we put m = 4 and in Table A2.2, m = 5. It will be observed that all the entries of Table A2.1 reappear in Table A2.2. This is a most important property of the orthogonal polynomials and the associated coefficients. It can be deduced from the arguments which we have developed for the computation of the $T_p(x)$ that this property is a necessary consequence of the orthogonality condition.

In addition, we calculated, but do not show here, all the elements of what we call the cross-check matrix. Each element say E_{pq}, is defined by

$$E_{pq} = \sum_i w_i T_p(x_i) T_q(x_i) .$$

In this case

$$E_{pq} = \sum_i T_p(x_i)T_q(x_i) .$$

If the orthogonality condition is satisfied, then all the elements off the diagonal should be zero. The values obtained were very small, of the order of 10^{-9}, and differed from zero due only to the rounding-off error of the computer.

2. Computation of $\{T_p(x)\}$, $p = 0,1,\ldots m$

We define these polynomials by the expression

$$T_p(x) = \sum_{q=0}^{p} f_{pq}x^q , \qquad f_{pp} = 1 , \tag{1}$$

which satisfy, as before, the orthogonality condition

$$\sum_i w_i T_p(x_i)T_q(x_i) = 0 , \quad p \neq q . \tag{2}$$

The first three polynomials are

$$T_0(x) = 1 ,$$

$$T_1(x) = f_{10} + x ,$$

$$T_2(x) = f_{20} + f_{21}x + x^2 .$$

To calculate polynomials of higher degree than two, we make use of the recurrence relation, equation (1.1.7), which applies equally here, and the simple result

$$f_{10} = -\Sigma w_i x_i/\Sigma w_i = -\bar{x} ,$$

which is shown in Chapter 14, Section 3. We proceed as before to build up the coefficients $\{g_{pq}\}$ and the polynomials $\{T_p(x_i)\}$.

A slight modification of our previous arguments is necessary to obtain the complete set of coefficients $\{f_{pq}\}$, $p,q = 0,1,\ldots m$. The subset $\{f_{p+1,0}\}$, $p = 1,2,\ldots m$, is obtained from the relation

$$f_{p+1,0} = -g_{pp}f_{p0} - g_{p,p-1}f_{p-1,0} \; . \tag{3}$$

Equation (3) is easily obtained from the recurrence relation, equation (1.1.7), by comparing coefficients of x^0. For the remainder of the coefficients, we proceed, as before, comparing coefficients of x_i^{q+1} in the expanded form of the recurrence relation to obtain

$$f_{p+1,q+1} = f_{pq} - g_{pp}f_{p,q+1} - g_{p,p-1}f_{p-1,q+1} \; . \tag{4}$$

In using equations (3) and (4), it is important to remember that

$$f_{pp} = 1, \quad \text{all } p$$

and

$$f_{pq} = 0, \quad q > p \; ,$$

the latter result following from the fact that there are no terms of higher power than x^p in the expansion of $T_p(x)$.

Table A2.3 shows the values of $\{T_p(x_i)\}$ for $p = 0,1,\ldots 5$ for $x_i = 0.5(0.5)5.0$ and $w_i = 1$, all i. Also given in this table are the numerical values of the coefficients $\{f_{pq}\}$ and $\{g_{pq}\}$. As before, the off-diagonal elements of the cross-check matrix were very small, differing from zero [the value required by equation (2)] only because of rounding-off error in the computer.

2.1. An alternative procedure when the x's are equally spaced and the points equally weighted

In this case, it is not necessary to calculate the orthogonal polynomials using the above procedures since quite extensive tabulations of their values are available for this case. It is usual to scale the x's so that they are symmetrically distributed about zero, using the linear transformation

$$u = (x - \bar{x})/\Delta x. \tag{1}$$

Here Δx is the spacing between adjacent x's and $\bar{x} = \Sigma x_i/n$. Clearly the set $\{u_i\}$ will consist of integers or half-integers (depending on whether n is odd or even) symmetrically distributed about zero.

It should be clear that the equations of Chapter 15 can be expressed in terms of u. For example, the equation of the fitted polynomial may be written

$$\hat{\mu} = \sum_{p=0}^{m} A_p T_p(u) \quad , \tag{2}$$

where, as before,

$$T_p(u) = \sum_{q=0}^{p} f_{pq} u^q \; . \tag{3}$$

However, because the set $\{u_i\}$ consists of equally-spaced integers or half-integers symmetrically distributed about zero, it can be shown that

$$T_p(-u) = (-1)^p T_p(u) \; . \tag{4}$$

Consequently

$$f_{pq} = 0$$

when p+q is odd. Thus, the terms containing odd powers of x are missing from the even T_p's and conversely even powers of x are missing from the odd T_p's. For example, the first four members of the set $\{T_p(u)\}$ are

$$T_0(u) = 1 \; ,$$

$$T_1(u) = u \; , \tag{5}$$

$$T_2(u) = u^2 - \Sigma u_i^2/n$$

$$= u^2 - (n^2 - 1)/12 \; , \tag{6}$$

$$T_3(u) = u^3 - (\Sigma u_i^4/\Sigma u_i^2)u$$

$$= u^3 - (3n^2 - 7)u/20 \; . \tag{7}$$

Guest (1961) gives formulae for the non-zero coefficients in terms of n for p = 1,2,...7.

A more convenient and extensive tabulation is that of Fisher and Yates (1957). These authors list the numerical values of the integers $T'_p(u)$, p = 1,2,...5, for n = 3,4,...75; $T'_p(u)$ is the integer derived from $T_p(u)$ by multiplication by a simple factor, λ_p, thus

$$T'_p(u) = \lambda_p T_p(u) \quad . \tag{8}$$

In terms of the T'_p's, equation (2) can be written

$$\hat{\mu} = \sum_{p=0}^{m} A'_p T'_p(u) \ , \tag{9}$$

where $A'_p = A_p/\lambda_p$. Thus, after scaling the x's to convert them into the u's, the whole calculation of the fitted polynomial can be performed using integers, with consequent computational simplicity if a desk machine is being used. The tables also include the numerical values of $\sum_i T'^2_p(u_i)$, p = 1,2,...5, to facilitate the calculation of the $\{A'_p\}$, and the values of λ_p so that the $\{A_p\}$ can be obtained. Incidentally, since $T_1(u) = u$, λ_1 is either 1 or 2 according to whether n is odd or even.

Example. To illustrate the simplicity of the Fisher and Yates tables, we have another look at the set of x values used in Table A2.3. There are 10 values in this set, viz. 0.5(0.5)5.0. First, we convert them into u's using equation (1) and obtain a set of half-integers −4.5(1.0)4.5. From the tables, we obtain the following values of $T'_p(u_i)$, λ_p, and $\sum_i T'^2_p(u_i)$, for p = 1 to 5. Note that it is not necessary to list the values of the T'_p's for the negative values of u since they may be obtained directly from the tabulated values using equation (4).

Obviously the tables may be used in the reverse way to obtain the values of $\{T_p(x)\}$ if these should be required for a given set of equally-spaced x's. The reader may verify that this is indeed the case by deriving the values of $\{T_p(x_i)\}$ shown in Table A2.3 (iii) from the values given below.

u_i	$T'_1(u_i)$	$T'_2(u_i)$	$T'_3(u_i)$	$T'_4(u_i)$	$T'_5(u_i)$
0.5	1	-4	-12	18	6
1.5	3	-3	-31	3	11
2.5	5	-1	-35	-17	1
3.5	7	2	-14	-22	-14
4.5	9	6	42	18	6
λ_p	2	1/2	5/3	5/12	1/10
$\sum_i T_p^2(u_i)$	330	132	8580	2860	780

Similar tabulations of orthogonal polynomials appear in Biometrika Tables and Owen's Handbook of Statistical Tables.

3. Scaling the x values

When polynomials of high degree, say $m \geq 5$, are to be fitted to experimental data, it will usually be found useful to rescale the x values so that the resulting transformed variable, x' say, lies in the interval (-2,2). A simple linear transformation is all that is required, viz.

$$x' = (x - a)/b,$$

where a and b are constants chosen to bring x' approximately into the above range. Working with the x''s, one is less likely to run into problems associated with the precision of the machine used for the computations. We have made similar remarks elsewhere without being quite so specific about the preferred range of the variable in question. It is outside the scope of this book to discuss the reasoning behind the above advice, but readers who wish to pursue the matter further are referred to Guest (1961) and Forsythe (1957) for further analysis of the problem.

Table A2.1 The case where $m = 4$ and $p, q > 0$

(i) The coefficients g_{pq}; $p = 1,2,3$; $q = 1,2,3,4$

3.2986	1		
1.0163	3.0293	1	
	1.3434	2.7767	1

(ii) The coefficients c_{pq}; $p, q = 1,2,\ldots 4$

1			
-3.9286	1		
10.884	-6.9578	1	
-24.945	28.861	-9.7345	1

(iii) The numerical values of the orthogonal polynomials

i	x_i	$T_1(x_i)$	$T_2(x_i)$	$T_3(x_i)$	$T_4(x_i)$
1	0.5	0.5	-1.7143	3.8277	-6.4115
2	1.0	1.0	-2.9286	4.9265	-4.8186
3	1.5	1.5	-3.6429	4.0464	-0.2721
4	2.0	2.0	-3.8571	1.9373	3.6770
5	2.5	2.5	-3.5714	-0.6506	4.9779
6	3.0	3.0	-2.7857	-2.9675	3.0796
7	3.5	3.5	-1.5000	-4.2633	-1.0686
8	4.0	4.0	0.2857	-3.7880	-5.0177
9	4.5	4.5	2.5714	-0.7916	-4.8186
10	5.0	5.0	5.3571	5.4759	4.9779

Note Where no entry is shown in the above grids, the value of the corresponding element is zero.

Table A2.2 The case when $m = 5$ and $p, q > 0$

(i) The coefficients g_{pq}; $p = 1,2,3,4$; $q = 1,2,3,4,5$

3.2986	1			
1.0163	3.0293	1		
	1.3434	2.7767	1	
		1.4192	2.6510	1

(ii) The coefficients c_{pq}; $p, q = 1,2,\ldots 5$

1				
-3.9286	1			
10.884	-6.9578	1		
-24.945	28.861	-9.7345	1	
50.681	-91.579	53.247	-12.385	1

(iii) The numerical values of the orthogonal polynomials

i	x_i	$T_1(x_i)$	$T_2(x_i)$	$T_3(x_i)$	$T_4(x_i)$	$T_5(x_i)$
1	0.5	0.5	-1.7143	3.8277	-6.4115	8.3588
2	1.0	1.0	-2.9286	4.9265	-4.8186	-0.9637
3	1.5	1.5	-3.6429	4.0464	-0.2721	-5.4294
4	2.0	2.0	-3.8571	1.9373	3.6770	-5.1431
5	2.5	2.5	-3.5714	-0.6506	4.9779	0.1718
6	3.0	3.0	-2.7857	-2.9675	3.0796	5.2863
7	3.5	3.5	-1.5000	-4.2633	-1.0686	5.1431
8	4.0	4.0	0.2857	-3.7880	-5.0177	-1.3931
9	4.5	4.5	2.5714	-0.7916	-4.8186	-7.7863
10	5.0	5.0	5.3571	5.4750	4.9779	3.9218

Note Where no entry is shown in the above grids, the value of the corresponding element is zero.

Table A2.3 The case when m = 5

(i) The coefficients g_{pq}; p = 0,1...4; q = 0,1,...5

2.7500	1				
2.0625	2.7500	1			
	1.6000	2.7500	1		
		1.4625	2.7500	1	
			1.3333	2.7500	1

(ii) The coefficients f_{pq}; p, q = 0,1,...5

1					
-2.750	1				
5.500	-5.500	1			
-10.725	19.025	-8.250	1		
21.450	-55.000	40.250	-11.000	1	
-44.687	147.33	-154.69	69.167	-13.750	1

(iii) The numerical values of the orthogonal polynomials

i	x_i	$T_0(x_i)$	$T_1(x_i)$	$T_2(x_i)$	$T_3(x_i)$	$T_4(x_i)$	$T_5(x_i)$
1	0.5	1	-2.25	3.0	-3.150	2.70	-1.8750
2	1.0	1	-1.75	1.0	1.050	-3.30	4.3750
3	1.5	1	-1.25	-0.5	2.625	-2.55	-0.3125
4	2.0	1	-0.75	-1.5	2.325	0.45	-3.4375
5	2.5	1	-0.25	-2.0	0.900	2.70	-1.8750
6	3.0	1	0.25	-2.0	-0.900	2.70	1.8750
7	3.5	1	0.75	-1.5	-2.325	0.45	3.4375
8	4.0	1	1.25	-0.5	-2.625	-2.55	0.3125
9	4.5	1	1.75	1.0	-1.050	-3.30	-4.3750
10	5.0	1	2.25	3.0	3.150	2.70	1.8750

Notes

(a) When no entry is shown in the above grids, the value of the corresponding element is zero.

(b) Much of the simplicity of sections (i) and (ii) arises because the x values are equally spaced. (See Section 2.1).

REFERENCES

ASPIN, A.A., 1948. An examination and further development of a formula occurring in the problem of comparing two mean values. Biometrika, 35, 88-96.

ASPIN, A.A. and WELCH, B.L., 1949. Tables for use in comparisons where accuracy involves two variances, separately estimated. Biometrika, 36, 290-296.

COLLETT, T. and LEWIS, T., 1976. The subjective nature of outlier rejection procedures. Appl. Statist., 25, 228-237.

DOWNHAM, D.Y. and ROBERTS, F.D.K., 1967. Multiplicative congruential pseudo-random number generators. Comp. J., 10, No.1, 74-77.

FERGUSON, T.S., 1961. On the rejection of outliers. Proc. 4th Berkeley Symp. Math. Statist. and Prob., 1, 253-287.

FISHER, R.A. and YATES, F., 1957. Statistical tables for biological, agricultural, and medical research, 5th edition (Oliver and Boyd).

FORSYTHE, G.E., 1957. Generation and use of orthogonal polynomials for data-fitting with a digital computer. J. Soc. Indust. Appl. Math., 5, 74.

FREEMAN, M.F. and TUKEY, J.W., 1950. Transformations related to the angular and square root. Ann. Math. Statist., 21, 607.

GUEST, P.G., 1961. Numerical methods of curve fitting. (Cambridge University Press).

LEVENE, H., 1960. Robust tests for equality of variances. (Contribution to Probability and Statistics, Stanford University Press, pp. 278-292).

OWEN, D.B., 1962. Handbook of statistical tables. (Pergamon Press).

PEARSON, E.S. and HARTLEY, H.O., 1958. Biometrika tables for statisticians. Vol. 1, 2nd edition. (Cambridge University Press).

PLACKETT, R.L., 1964. The continuity correction in 2 x 2 tables. Biometrika, 51, 327-337.

REILLY, P.E.B. and FORD, E.J.H., 1974. The effect of betamethasone on glucose production and on gluconeogenesis from amino acids in sheep. J. Endocr., 60, 455.

SATTERTHWAITE, F.E., 1946. An approximate distribution of estimates of variance components. Biometrics, 4, 110-114.

WELCH, B.L., 1947. The generalization of 'Student's' problem when several different population variances are involved. Biometrika, 34, 28-35.

WELFORD, B.P., 1962. Note on a method for calculating corrected sums of squares and products. Technometrics, 4, 419.

INDEX